AF531650

PRODUCTION DISEASES OF DAIRY ANIMALS

PRODUCTION DISEASES OF DAIRY ANIMALS

Dr. V.K. Singh

RANDOM PUBLICATIONS

NEW DELHI - 110 002 (INDIA)

Production Diseases of Dairy Animals

ISBN 978-93-51116-88-2

Published in 2015 in India by

RANDOM PUBLICATIONS

4376-A/4B, Gali Murari Lal, Ansari Road
New Delhi-110 002
Phone: +9111-43580356, 23289044
E-mail: randomexports@gmail.com; sales@randompublications.com; info@randompublications.com

Reprinted 2025

Type Setting by: Friends Media, Delhi-110089
Digitally Printed at: Replika Press Pvt. Ltd.

Preface

Dairy cattle are susceptible to the same diseases as beef cattle. Many diseases and pests plague the cattle industries of the world, the more serious ones being prevalent in the humid and less developed countries. One of the more common diseases to be found in the developed countries is brucellosis, which has been controlled quite successfully through vaccination and testing. This disease produces undulant fever in humans through milk from infected cows. Leptospirosis, prevalent in warm-blooded animals and humans, is caused by a spirochete and results in fever, loss of weight, and abortion. Bovine tuberculosis has been largely eliminated; where it has not, it can infect other warm-blooded animals, including humans. Test and slaughter programmes have proved effective. Rabies, caused by a specific virus that also can infect most warm-blooded animals, is usually transmitted through the bite of infected animals, either wild or domestic. Foot-and-mouth disease has been eliminated from most of North America, some Central American countries, Australia, and New Zealand. The rest of the world is still plagued by the disease, which attacks all cloven-footed animals. Humans are mildly susceptible to this organism. Successful vaccinations have been developed for blackleg, malignant edema, infectious bovine rhinotracheitis (or red nose), and several other diseases. Anaplasmosis, common to most tropical and semitropical regions, is spread by the bite of mosquitoes and flies. Anthrax, caused by a generally fatal bacterial infection, has been largely eliminated in the United States and western Europe. Rinderpest, still common to Asia and Europe, is caused by a specific virus that produces high fever and diarrhea. An infectious fever sometimes called nagana, caused by the tsetse fly, attacks both cattle and horses and is prevalent in central and southern Africa, as well as in the Philippines. Grass tetany and milk fever both result from metabolic disturbances. Bloat, caused by rapid gas formation in the rumen, or first compartment of the stomach, is sometimes fatal unless relieved. Pinkeye is an infectious inflammation of the eyes spread by flies or dust and is most serious in cattle having white pigmentation around one or both eyes. Mastitis, an inflammation of the udder, is caused by rough handling or by infection. Vibriosis, a venereal disease that causes abortion; pneumonia, an inflammation of the lungs; and shipping fever all cause serious losses and are difficult to control except through good management. Broad-spectrum antibiotics (antibiotics that are effective against various microorganisms), as well as powerful and specific pharmaceuticals, are effective and profitable means of keeping cattle herds healthy. Vermifuges, which destroy or expel parasitic worms, and insecticides, which kill harmful insects, are also highly effective and much used.

Pigs are relatively easy to raise indoors or outdoors, and they can be slaughtered with a minimum of equipment because of their moderate size. Pigs are monogastric, so, unlike ruminants, they are unable to utilize large quantities of forage and must be given concentrate feed. Furthermore, pigs have only one primary economic use—as a source of meat (pork) and lard—unlike most other livestock, such as cattle and sheep, which have many other important economic uses. An important factor that influences dairy herd productivity is the amount and type of disease in the herd. The basis of disease control programmes includes knowledge of the frequency of disease, information about the biologic effect of disease, and information on the effectiveness of control procedures. Increased culling, reduced milk or protein yield, increased adult cow mortality, and reduced reproductive efficiency are all potential results of disease in adult cows. Milk production is often profoundly reduced in cows with clinical disease. Being able to properly identify and monitor certain health conditions and diseases of dairy cattle is an important step in preventing side effects from these situations.

The present book is the essential reference for practising veterinarians, herd health practitioners, extension officers and other farm advisors, as well as dairy farmers.

I thank all members of my team who have helped in the preparation of the book. My special thanks go to "Random Publications" who have published the book.

— *Dr. V.K. Singh*

Contents

Chapter 1

Diseases from Farm Animals

Farm animals including cows, sheep, pigs, chickens and goats, can pass diseases to people. As you know, farm animals are not like house pets and do not have places to rest or eat that are away from where they pass manure. Therefore, you should thoroughly wash your hands with running water and soap after contact with them or after touching things such as fences, buckets, and straw bedding, that have been in contact with farm animals, adults should carefully watch children who are visiting farms and help them wash their hands well.

Different types of farm animals can carry different diseases. For example, cows and calves can carry the bacterium *Escherichia coli* 0157:H7, often called *E. coli* (ee COH-lie). This germ can cause bloody diarrhea in people. In addition children can develop kidney failure due to *E. coli* 0157:H7 infection. Pigs can carry the bacterium *Yersinia enterocolitica* (yer-SIN-ee-ah en-TER-o-koh-LIH-tee-kuh), which causes the disease yersiniosis (yer-SIN-ee-OH-sis). Chickens can carry bacteria such as Salmonella, (sal –mon – Nell – ah) which causes the disease salmonellosis. Many of these germs are in farm animal manure.

Some people are more likely than others to get diseases from farm animals. A person's age and health status may affect his or her immune system, increasing the chances of getting sick. People who are more likely to get diseases from farm animals include infants, children younger than 5 years old, organ transplant patients, people with HIV/ AIDS, and people who are being treated for cancer. Special advice is available for people who are at greater risk than others of getting diseases from animals.

BSE (Bovine Spongiform Encephalopathy, or Mad Cow Disease)

BSE (bovine spongiform encephalopathy) is a progressive neurological disorder of cattle that results from infection by an unusual transmissible agent called a prion. The nature of the transmissible agent is not well understood. Currently, the most accepted theory is that the agent is a modified form of a normal protein known as prion protein. For reasons that are not yet understood, the normal prion protein changes into a pathogenic (harmful) form that then damages the central nervous system of cattle.

Research indicates that the first probable infections of BSE in cows occurred during the 1970's with two cases of BSE being identified in 1986. BSE possibly originated as a result of feeding cattle meat-and-bone meal that contained BSE-infected products from a spontaneously occurring case of BSE or scrapie-infected sheep products. Scrapie is a prion disease of sheep. There is strong evidence and general agreement that the outbreak was then amplified and spread throughout the United Kingdom cattle industry by feeding rendered, prion-infected, bovine meat-and-bone meal to young calves.

The BSE epizootic in the United Kingdom peaked in January 1993 at almost 1,000 new cases per week. Over the next 17 years, the annual numbers of BSE cases has dropped sharply; 14,562 cases in 1995, 1,443 in 2000, 225 in 2005 and 11 cases in 2010. Cumulatively, through the end of 2010, more than 184,500 cases of BSE had been confirmed in the United Kingdom alone in more than 35,000 herds.

There exists strong epidemiologic and laboratory evidence for a causal association between a new human prion disease called variant Creutzfeldt-Jakob disease (vCJD) that was first reported from the United Kingdom in 1996 and the BSE outbreak in cattle. The interval between the most likely period for the initial extended exposure of the population to potentially BSE-contaminated food (1984-1986) and the onset of initial variant CJD cases (1994-1996) is consistent with known incubation periods for the human forms of prion disease.

BSE (bovine spongiform encephalopathy) is a progressive neurological disorder of cattle that results from infection by an unusual transmissible agent called a prion. The nature of the transmissible agent is not well understood. Currently, the most accepted theory is that the agent is a modified form of a normal protein known as prion protein. For reasons that are not yet understood, the normal prion protein changes into a pathogenic (harmful) form that then damages the central nervous system of cattle.

Research indicates that the first probable infections of BSE in cows occurred during the 1970's with two cases of BSE being identified in 1986. BSE possibly originated as a result of feeding cattle meat-and-bone meal that contained BSE-infected products from a spontaneously occurring case of BSE or scrapie-infected sheep products. Scrapie is a prion disease of sheep. There is strong evidence and general agreement that the outbreak was then amplified and spread throughout the United Kingdom cattle industry by feeding rendered, prion-infected, bovine meat-and-bone meal to young calves.

The BSE epizootic in the United Kingdom peaked in January 1993 at almost 1,000 new cases per week. Over the next 17 years, the annual numbers of BSE cases has dropped sharply; 14,562 cases in 1995, 1,443 in 2000, 225 in 2005 and 11 cases in 2010. Cumulatively, through the end of 2010, more than 184,500 cases of BSE had been confirmed in the United Kingdom alone in more than 35,000 herds.

There exists strong epidemiologic and laboratory evidence for a causal association between a new human prion disease called variant Creutzfeldt-Jakob disease (vCJD) that was first reported from the United Kingdom in 1996 and the BSE outbreak in cattle.

The interval between the most likely period for the initial extended exposure of the population to potentially BSE-contaminated food (1984-1986) and the onset of initial variant CJD cases (1994-1996) is consistent with known incubation periods for the human forms of prion disease.

BSE (bovine spongiform encephalopathy) is a progressive neurological disorder of cattle that results from infection by an unusual transmissible agent called a prion. The nature of the transmissible agent is not well understood. Currently, the most accepted theory is that the agent is a modified form of a normal protein known as prion protein. For reasons that are not yet understood, the normal prion protein changes into a pathogenic (harmful) form that then damages the central nervous system of cattle.

Research indicates that the first probable infections of BSE in cows occurred during the 1970's with two cases of BSE being identified in 1986. BSE possibly originated as a result of feeding cattle meat-and-bone meal that contained BSE-infected products from a spontaneously occurring case of BSE or scrapie-infected sheep products. Scrapie is a prion disease of sheep. There is strong evidence and general agreement that the outbreak was then amplified and spread throughout the United Kingdom cattle industry by feeding rendered, prion-infected, bovine meat-and-bone meal to young calves.

The BSE epizootic in the United Kingdom peaked in January 1993 at almost 1,000 new cases per week. Over the next 17 years, the annual numbers of BSE cases has dropped sharply; 14,562 cases in 1995, 1,443 in 2000, 225 in 2005 and 11 cases in 2010. Cumulatively, through the end of 2010, more than 184,500 cases of BSE had been confirmed in the United Kingdom alone in more than 35,000 herds.

There exists strong epidemiologic and laboratory evidence for a causal association between a new human prion disease called variant Creutzfeldt-Jakob disease (vCJD) that was first reported from the United Kingdom in 1996 and the BSE outbreak in cattle. The interval between the most likely period for the initial extended exposure of the population to potentially BSE-contaminated food (1984-1986) and the onset of initial variant CJD cases (1994-1996) is consistent with known incubation periods for the human forms of prion disease.

Brucellosis

Clinical Features In the acute form (<8 weeks from illness onset), nonspecific and "flu-like" symptoms including fever, sweats, malaise, anorexia, headache, myalgia, and back pain. In the undulant form (<1 year from illness onset), symptoms include undulant fevers, arthritis, and epididymo-orchitis in males. Neurologic symptoms may occur acutely in up to 5% of cases.

In the chronic form (>1 year from onset), symptoms may include chronic fatigue syndrome, depression, and arthritis. Etiologic Agent *Brucella* species, usually *B. abortus* (cattle), *B. melitensis, B.ovis* (sheep, and goats), *B. suis* (pigs), and rarely *B. canis* (dogs). Incidence In the United States, < 0.5 cases per 100,000 population, primarily *B. melitensis*. . Most cases are reported from California, Florida, Texas, and Virginia. Sequelae Variable, including granulomatous hepatitis, peripheral arthritis, spondylitis, anemia, leukopenia, thrombocytopenia, meningitis, uveitis, optic neuritis, papilledema, and endocarditis.

Transmission Zoonotic. Commonly transmitted through abrasions of the skin from handling infected mammals. In the United States, occurs more frequently by ingesting unpasteurised milk or dairy products. Highly infectious in the laboratory via aerosolisation; handling cultures warrants biosafety level-3 precautions. Risk Groups Abattoir workers, meat inspectors, animal handlers, veterinarians, and laboratorians. Surveillance Brucellosis is a nationally notifiable disease and reportable to the local health authority. Trends For previous 10 years, approximately 100 cases per year have been reported. Challenges Elimination of domestic and feral animal reservoirs. In 2001, the National Brucellosis Eradication Program reported only 3 newly affected cattle herds, compared to 14 herds identified in 2000. Establish and validate methods for isolation and detection of *Brucella* spp. in foods.

Campylobacter Infection and Animals

Campylobacteriosis is a bacterial disease caused by *Campylobacter jejuni* or *Campylobacter coli. Campylobacter* usually causes a mild to severe infection of the gastrointestinal system, including watery or bloody diarrhea, fever, abdominal cramps, nausea, and vomiting. A rare complication of *Campylobacter* infection is Guillain-Barre syndrome, a nervous system disease that occurs approximately 2 weeks after the initial illness develops.

Can Animals Transmit Campylobacter *to Me?*

Sometimes, yes, animals can spread *Campylobacter* to humans. Most people get campylobacteriosis from contaminated food. However, animals can have *Campylobacter* in their feces (stool). If people touch contaminated feces, they can get sick. Animals that may carry *Campylobacter* in their feces include farm animals, cats, and dogs.

Animals do not have to be ill to pass *Campylobacter* to humans. People with compromised immune systems, including those undergoing treatments for cancer, organ transplant patients, and people with HIV/ AIDS, have a higher risk than others of getting *Campylobacter* infection from food and animals.

Campylobacter Jujuni

How do I reduce my risk of getting *Campylobacter* infection from animals?

- After contact with animals and animal feces, wash your hands thoroughly with running water and soap.
- If you are immunocompromised and are getting a new pet, avoid farm animals, cats, and dogs with diarrhea.
- If your dog or cat has diarrhea, talk to your veterinarian.
- If you develop symptoms, including diarrhea, vomiting, abdominal cramps, and/or nausea, contact your physician. Be sure to inform him or her of your pet and if it is ill.
- If you are immunocompromised, be extra cautious around farm animals and their environment.

Cryptosporidium Infection and Animals

What is Cryptosporidium *Infection?*

Cryptosporidium infection (cryptosporidiosis) (krip-toe-spo-rid-ee-oh-sis) is a parasitic disease caused by *Cryptosporidium parvum*. It usually causes a mild to severe infection of the gastrointestinal system, including watery diarrhea, fever, abdominal cramps, nausea, and vomiting.

Can Animals Give Me Cryptosporidium Infection?

Yes, sometimes. Most people get *Cryptosporidium* infection from contaminated food and water. However, sometimes animals (including farm animals, cats, and dogs) carry this parasite in their feces (stool) and pass it to people. Animals do not have to be ill to pass *Cryptosporidium* to humans. People with compromised immune systems, such as those undergoing immunosuppressive treatments for cancer, organ transplant patients, and people with HIV/AIDS, are more likely than others to get *Cryptospordium* infection.

How do I reduce my risk of acquiring cryptosporidiosis from my pet?

- After contact with animals and animal feces (stool), wash your hands thoroughly with running water and soap.
- If you are immunocoromised and are getting a new pet, avoid strays, puppies, kittens and pets with diarrhea.
- If your dog or cat has diarrhea, take it to your veterinarian.
- If you develop symptoms, including diarrhea, vomiting, abdominal cramps, and/or nausea, contact your physician. Be sure to inform him or her of your pet and if it is ill.
- If you are immunocompromised, be extra cautious around farm animals and their environment.

How do I find more information on *Cryptosporidium* infection?

To learn more about this disease, refer to CDC's site on *Cryptosporidium* infection, including fact sheets, prevention tips, and other resources.

Escherichia Coli Infection and Farm Animals

What is Escherichia Coli O157 (E. coli)?

Escherichia coli O157 is a species of bacteria. The most common type of *E. coli* infection that causes illness in people is called *E. coli* O157. Symptoms of *E. coli* O157 include watery or bloody diarrhea, fever, abdominal cramps, nausea, and vomiting. Illness may be mild or severe. Young children are more likely to have severe symptoms, including kidney failure, and die.

How is E. Coli Transmitted from Animals to People?

While most people get *E. coli* O157 from contaminated food (such as undercooked ground beef), it also can be passed in the manure (feces) of young calves and other cattle. Animals do not have to be ill to transmit *E. coli* O157 to humans.

How I Do Reduce My Risk of Getting E. Coli from Animals?

- After contact with cattle or their manure (feces), wash your hands thoroughly with running water and soap.
- Children under the age of 5 years old should be extra cautious around cattle (including those in petting zoos).
- If you develop symptoms, including diarrhea, vomiting, abdominal cramps, and/or nausea, contact your physician. Be sure to inform him or her of recent contact with farm animals.

How Do I Find More Information about E. Coli?

Please read CDC's *E. coli* fact sheet for answers to frequently asked questions, technical information, and references. In addition, refer to the articles below for information about *E. coli* infection at petting zoos and farms and in swimming pools.

Yersinia Enterocolitica and Pigs

What is Yersiniosis?

Yersiniosis (yer-SIN-ee-O-sis) is a disease caused by the bacterium *Yersinia enterocolitica*. People with yersiniosis can have different symptoms depending on how old they are. People can start to get sick 4 to 7 days after infection and can be sick for 1 to 3 weeks. Young children usually have fever, stomach pain, and diarrhea. Adults do not get sick with yersiniosis as often, but they can feel pain on their right side and may have a fever. Usually, these signs go away after about 3 weeks but sometimes pain in joints, such as knees or wrists, can start after that and last for several months.

Can Animals Transmit Yersiniosis to Me?

Yes, some animals pass *Yersinia enterocolitica* in their feces (stool) and people can get sick from contact with infected feces. Several kinds of animals can carry this disease, but usually people get sick from pigs that are sick with yersiniosis. Other animals that can carry this disease include cats, dogs, horses, cows, rodents, and rabbits. People can also get yersiniosis by eating pork that is not cooked completely or by drinking contaminated milk.

How Can I Protect Myself from Getting Yersiniosis?

- Avoid eating raw or undercooked pork.
- Consume only pasteurised milk or milk products.

- Wash hands with soap and running water before eating and preparing food, after contact with animals, and after handling raw meat.
- After handling raw chitterlings (food prepared from small intestine of pigs), clean hands and fingernails thoroughly with soap and water before touching infants or their toys, bottles, or pacifiers. Someone other than the foodhandler should care for children while chitterlings are being prepared.
- Prevent cross-contamination in the kitchen: Use separate cutting boards for meat and other foods, and carefully clean all cutting boards, countertops, and utensils with soap and hot water after preparing raw meat.
- Dispose of animal feces in a sanitary manner.

Infectious Cattle Diseases and Vaccines

Vaccines are available for 20 to 30 infectious diseases of cattle. With the various brand names and different combinations available, the choice of vaccines can become very complicated. Calves vaccinated under 6 months of age should generally be re-vaccinated again after that age to provide a longer lasting immunity. It is important to follow the specific directions provided with a vaccine. If two doses are recommended initially, don't count on very much protection until 7–14 days after the second dose has been given. The major diseases for which vaccines are available are categorised and briefly described below.

Sudden Death

Clostridial Diseases: These diseases include Blackleg, Malignant Edema, Black's Disease, Enterotoxemia and Redwater. All of these are common diseases. The organisms form spores that may survive a long time in hostile environments and yet kill cattle quickly, giving little opportunity for treatment. The vaccines produce good immunity, but most require two doses initially to really be effective. Some producers give only one Blackleg vaccination to calves and have no losses.

Because of the sporadic nature of these diseases, this lack of loss is probably due to a lack of exposure rather than a high level of immunity. With the other clostridial vaccines, the second dose is essential for stimulation of protective immunity (there is one designed as a one-dose vaccine).

Anthrax: This cause of sudden death has occurred in at least three areas in Utah, but is only seen sporadically. The organism will survive indefinitely in the soil and when conditions are right, multiply and cause a disease outbreak. A vaccine is available, but should only be used when cattle are grazed in known problem areas.

Respiratory Diseases

IBR (Infectious Bovine Rhinotracheitis) (Rednose): A viral infection of the upper respiratory tract. It is present in almost all herds, but causes illness in unexposed animals or those with lowered levels of immunity. Many cattle carry the virus and shed it to others during periods of stress. This agent is commonly implicated with bacterial agents in causing shipping fever and extension.usu.edu other severe cases of pneumonia.

Both MLV (modified live virus) vaccines and killed (or attenuated) products are available. Some are designed for IM (intramuscular) use while others are given IN (intranasally). The killed and intranasal products may be used in, or around, pregnant cows but other vaccines may cause abortions. The IN vaccines will cause some antibody response within three days and may be useful even in the face of an outbreak. Two doses of a killed product must be used to confer protective immunity. All replacement animals should be vaccinated. For intensively managed herds, annual boosters are recommended.

PI3 (Parainfluenza-3): Another viral respiratory agent that causes a relatively mild disease by itself, but a severe problem when combined with a bacterial agent. It is included with most IBR vaccines and can be used on the same schedule.

BVD (Bovine Virus Diarrhea): A common viral agent, present in almost all herds. It may cause respiratory, digestive tract or reproductive problems. It has a profound detrimental affect on the immune system.

A number of MLV vaccines have been available. Killed vaccines are also available which stimulate a good immune response in adult animals. But, they are apparently not able to protect the fetus. Two doses are required initially. All replacement animals should be vaccinated with an MLV vaccine (perhaps even two doses), after 6 months of age and prior to breeding.

BRSV (Bovine Respiratory Syncytial Virus): A relatively recently recognised disease agent, but now identified all across the country in respiratory infections. It is mainly a problem in weaned and feedlot animals (also young dairy stock). Both MLV and killed virus vaccines are available with two initial doses required for both. Pasteurella: A bacteria carried by many normal cattle. It becomes a major cause of severe "shipping fever" pneumonia when combined with stress and a viral agent. Two species are common: P. hemolytica and P. multocida. Vaccines available in the past were poor, with use of a single dose causing more problems than if none were used. Great improvements have been made in recent years and several newer products are available, with more to come. Both one and two dose products are now available. Follow directions carefully for these products to be beneficial. They must usually be given prior to weaning in order to help hold down the occurrence of disease at this critical time.

Haemophilus Sommus: This agent is the other major bacterial agent involved in shipping fever. It also causes "brain fever" in feedlot cattle (also known as TEME: thromboembolic meningioencephalitis). The killed vaccine must be given in two doses initially and should be used prior to weaning for the greatest benefit.

Reproductive Diseases

IBR: The most common cause of abortion in cattle. All replacements should be vaccinated to protect against it. Use the proper vaccine for, or around, pregnant cows.

BVD: May cause early embryonic death, abortion or congenital defects. Vaccinate all replacement animals with a MLV vaccine (two doses preferred) to produce a planned exposure between 6 months of age and breeding.

Brucellosis (Bangs): Many states are free of this disease agent, but vaccination is still recommended (and required for sale) until the threat from other states is further reduced. All heifers 4–12 months of age should be vaccinated by an accredited veterinarian. They must also be ear tagged and tattooed. Many states will not accept animals for breeding purposes unless the tattoo can be identified.

Vibriosis (Campylobacter): A common bacterial disease, spread through breeding. It causes early embryonic death so appears as an infertility and results in a prolonged breeding and calving season as well as a reduced calf crop. Two types of vaccine are available. One is in an oil base product to prolong the absorption. Only one dose is required initially. Subsequent boosters given in the fall at pregnancy testing will extend the protection on through the next breeding season.

The other type of vaccine has an aluminium hydroxide or other adjuvant and requires two doses initially. Be sure to give both doses to obtain a protective level of immunity. The annual boosters for this type of vaccine should be given 30 days prior to breeding. This type comes in combination with Lepto and other vaccines and is easier to administer, but it must be used according to directions if it is to be effective.

Bulls infected with vibrio have been cleared by use of two doses (of 5 ml) of the oil base vaccine, 30 days apart. All bulls should be vaccinated and given an annual booster in the fall. All cows in multiple owner herds and in herds adding used cows or bulls should be vaccinated. Lepto: May cause abortion and illness. It has not been commonly diagnosed in Utah in recent years, but is very difficult to diagnose. It is spread through urine and water contamination so is a potential threat to almost all cattle. It may also be carried by other species including rodents, dogs, swine and man.

Trichomoniasis: A common disease in Utah that causes early embryonic death. A vaccine is available with an efficacy of about 50%. It should be considered for use in infected herds which are mixed with other during the breeding season. Two doses are required initially. The second (and the annual booster) should be given 30 days prior to the beginning of breeding.

Scours

Rota and Corona Virus: Two viral agents that are common in Utah herds and contribute to scours. They usually produce only mild disease signs by themselves, but become more severe when combined with stress or other agents. A vaccine is available for problem herds. It requires two doses the first year and the last dose should be given at least three weeks prior to the start of calving. It may be used in combination with E. coli.

Coli (Coliform): A bacterial cause of scours that usually appears in calves under 5 days of age. A common contaminant in manure and may build up to epidemic levels. A vaccine requiring one or two initial doses is available for the cows prior to calving. A monoclonal antibody product is also available for use on calves at birth (by mouth) if a serious outbreak occurs in calves from dams that have not been vaccinated. It will provide some immediate, but short term protection.

Pinkeye

An eye infection due to a specific bacteria, that is commonly carried by many normal cattle. A herd outbreak is often precipitated by eye irritation (dust, sunlight, etc.). The infection is readily spread to other animals by face flies. Several vaccines are now available. Since the immunity they stimulate is not very long-lasting, these vaccines should be administered in the spring just prior to the fly season. This will provide the greatest protection during summer, the period of greatest exposure.

Pathogenic Bacteria

Bacteria that cause disease are called pathogenic bacteria. Bacteria can cause diseases in humans, in other animals, and also in plants.

Some bacteria can only make one particular host ill; others cause trouble in a number of hosts, depending on the host specificity of the bacteria. The diseases caused by bacteria are almost as diverse as the bugs themselves and include food poisoning, tooth ache anthrax, even certain forms of cancer. It is impossible to sum up all bacterial diseases and it would be pretty boring. The Infectious Diseases fact sheets gives brief descriptions of diseases, including infectious diseases. Some diseases are named after the organisms that cause them, or is it the other way round?

If you want to have a look at pathogenic bacteria under the microscope:

- View here pathogenic bacteria under the microscope which also describes their properties, as part of a course on medical microbiology. Some of the terms used here are explained in our exhibit on pathogenicity.
- More links to microscopic pictures can be found in the exhibit on Images of bacteria.

Some pathogenic bacteria have received disproportionate attention in the press, e.g. the 'flesh-eating bacteria', which in real life are called Streptococci. Indeed they can cause spectacular, but fortunately uncommon symptoms. In the press, pathogenic bacteria are sometimes represented as (deadly) dangerous enemies that lurk in the dark, unseen, ready to attack you. Although that is exaggerated some bacteria can be life treatening, for example *Legionella pneumoniae*, the causative of Legionnaire's disease. These bacteria survive in moist places like air conditioners or hot-water pipes.

Though potentially life treatening, bacterial infections do not kill all their victims. Some bacteria kill a high percentage of people infected (they have a so-called a high mortality rate), but their relatively inefficient rate of spreading makes up for that. Other bacteria spread very easily, but they don't kill many of the people they infect. Imagine what would happen with a pathogen that spread very fast, killed all its victims; it would soon have none left to infect, and would die out together with the species on which it had been dependent. Bacterial pathogens don't eradicate their hosts completely. Although a popular theme for thrillers, this is a 'mission impossible' in real life.

When an infectious disease spreads around an area, and the cases of new infections reach a certain number, we call it an epidemic. Read about epidemics or visit the Epidemics Home Site. If, on the other hand, a certain disease is always present in low numbers of cases in a given area, that disease is said to be endemic in that area. Some epidemics become wide-spread and quickly reach distant parts of the world: in modern times people travel fast and frequent, and our bacteria travel with us. A classical example of such a pandemic (in this case caused by a virus) was a new type of influenza ("Spanish 'flu") that reached continent after continent early twentieth century, killing thousands of people on it's way. Although only a small proportion of the infected people died (the virus had a low mortality) so many people got infected that even the small proportion of deaths amounted to large numbers.

An epidemic or pandemic can only occur if the population is not immune to that disease. Read our special feature file on the history of infectious diseases about the times that pathogens could spread unlimited. Our exhibit on our immune system explains how immunity can prevent disease.

So why do epidemics occur? Either because they are caused by diseases that did not exist before, like AIDS, caused by the HIV virus, or because new variants of bacteria (or viruses) arrive in an area where they were not endemic before. This is why epidemics of the common 'flu' occur frequently: the virus causing influenza is able to change itself sufficiently to bypass our immunity built up from prior infections. Although not caused by bacteria, this site on influenza is very interesting if you want to understand epidemics of infectious diseases. Whenever a new virus type arises, there is the potential for a new epidemic. Fortunately, most of these new types are not as vicious as the Spanish 'flu. The experts are currently keeping an eye on bird flu, a strain of influenza that has not (yet?) learned to jump from person-to-person. If that would happen, a new pandemic could be the result.

An example of a bacterial disease that caused successive epidemics, and even pandemics, in recent times is *Vibrio cholerae*, the cause of cholera. Epidemiology is the study to establish the cause of a disease.

- Read this classic of cholera in London in the last century: an epidemiological 'whodunit' with John Snow as the chief inspector.
- Another classic epidemic in history was small pox raging through Europe in the 14th Century. Read our special feature file on the Black Death.

Bacteria have invented many different strategies to make us ill. These strategies, called bacterial pathogenicity, are the subject of an important division of medical microbiology. Understanding how certain bacteria make us ill can result in better treatment, vaccination, or prevention of that infectious disease. In another exhibit some of the common mechanisms of bacterial pathogenicity are explained. In order to keep this information balanced, now that you know what pathogenic bacteria are, why not also check out how commensal bacteria are good for you. You take healthy animals and often very quickly after you vaccinate, you can see simple things like itching of the skin or excessive licking of the paws, sometimes even with no eruptions. We see a lot of epilepsy, often after a rabies vaccination. Or dogs or cats can become aggressive for several days. Frequently, you'll see urinary tract infections in cats, often within three months after their [annual] vaccination. If you step back, open your mind and heart, you'll start to see patterns of illness post-vaccination."

Bloat and Torsion

"Bloat" or torsion of the stomach is sudden and often fatal. It can occur to any healthy large breed dog. Bloat is a condition where the stomach organ has twisted, cutting off vital functioning. It only happens in big dogs. The disease of bloat and torsion manifests itself under stressful conditions: vaccine, drug, anesthesia reactions and commercial kibble or canned food.

My German Shepheard dog, Kuuma, gets hit by this "bloat", or stomach spams once a month or so. It usually hits him soon after he eats kibbles (never when he eats only raw food - I feed him rawfood and kibbles because 100% raw gives him diarrheat) . Occasionally after he eats, he wakes up extremely agitated and restless, and tries to nibble on just about anything. It is obvious that he's in great pain. One thing that helps him, is to let him eat plenty of grass. He'll be grazing for 20 min or longer non-stop. If no grass is available, I let him eat wheatgrass which I grow in trays by the windowsill (I drink wheatgrass juice everyday). I have step by step instructions how to grow wheatgrass in apartments or homes. I believe that Kuuma would actually die if no grass was available for him to eat.

"Tragic middle-of-the-night trip to the emergency clinic, stomach the size of a basketball, turned upside-down. All they found in his stomach was kibble. I had suspected for a long time that something was wrong with his digestion. His stomach was always so noisy, he was always burping and farting. And the diarrhea! He was unable to digest and absorb commercial food. The kibble connection is a strong factor in bloat cases. He is now on a completely raw, whole food diet (BARF for Bones And Raw Foods). I've done a great deal of investigating, and have found out that it is VERY uncommon for dogs on the raw diets to bloat.

As far as I can tell, the best prevention for bloat is a raw whole food diet, done properly, of course. There are countless other benefits of the raw diet as well. There are many problems with the cooked diets that most people feed their dogs— the first being that the food is *cooked* and any enzymes have been killed, along with much nutritional value. Then there's the whole issue of very suspicious ingredients in commercial foods, significant toxins, and the fact that it's all got a very high percentage of grain. And the fats, once they're heated, turn into harmful trans fatty acids."

Nutrition: Is it Factor in Bloat and Torsion?

"It has been my experience that the number of incidents of bloat/torsion have dropped dramatically over the past few years probably due to better quality meat based foods and the incorporation of whole foods, probiotics and digestive enzymes into the diet. I base this comment on the fact that I have had very few calls regarding bloat and torsion in 4+ years. Before that, I would average 2-3 calls for assistance per week. It is because of this experience and my interest in the prevention of disease through quality nutrition that has made me consider taking a closer look at the effects diet can have on the cause and prevention of bloat and torsion."

Leaky Gut Syndrome and Toxic Gut Syndrome can be the silent killers behind the cause for Bloat/Torsion and the misery of allergies. In this article I would like to explore what happens when a system is overrun with toxic levels of fungus, also known as yeast and high levels of pathogenic (bad) bacteria. Candida Albicans is a Fungus/Yeast and a common microorganism that lives in the gut. But when there is an "overgrowth" of this fungus/yeast in the gut, it is called a systemic yeast infection, and it affects the health and well being of the whole animal or human. When on antibiotics or stressed, the pH balance of the gut is out of balance, and beneficial bacteria in the gut may be destroyed, and this insidious fungus and/or other pathogenic bacteria can take over and this overgrowth is very detrimental to our health and well being.

Sissy McGill knew many dog owners who had experienced the bloat problem in their animals. Sissy became interested in pet nutrition in 1974 after three of her Great Danes died of "bloat," a canine digestive problem. She learned that "bloat" was a common cause of death for dogs in the USA, but that the German vocabulary did not even assign a word for "bloat." Additionally, Tufts University research at the time revealed that "only in the USA do dogs die of bloat." It also suspected that the reason could be the corn, wheat, and soy in their diet. [Ed. Note – Interesting, since even people can get bloat from corn, wheat, and soy.] Europeans feed their dogs a natural type of food that doesn't contain the animal fat and soybean products.

In 1975, Sissy introduced the first all-natural dog food in the USA, containing no salt, animal fat, sugar-beet pulp, fillers, by-products, artificial flavours, colours, preservatives, corn, soy, or wheat – the last three being the top allergens in dogs. Learn how the Food and Drug Administration (FDA) relentlessly harassed her for selling animal products made from all natural ingredients. She ended up in jail. * Naturally relieve unpleasant symptoms of flatulence and gas in pets:

* Reduce bloating, abdominal cramps and colic
* Maintain digestive health and help with the symptoms of digestive disorders, including diarrhea and constipation.

Veterinary Medicine

Several diet-related factors were associated with a higher incidence of bloat. These include feeding only dry food, or feeding a single large daily meal. Dogs fed dry foods containing fat among the first four ingredients had a 170 percent higher risk for developing bloat. Dogs fed dry foods containing citric acid and were moistened prior to feeding had a 320 percent higher risk for developing bloat. Conversely, feeding a dry food containing a rendered meat-and-bone meal decreased risk by 53 percent in comparison with the overall risk for the dogs in the study. Mixing table food or canned food into dry food also decreased the risk of bloat.

During the past 30 years there has been a 1,500 percent increase in the incidence of bloat, and this has coincided with the increased feeding of dry dog foods. There is a much lower incidence of bloat in susceptible breeds in Australia and New Zealand. Feeding practices in these countries have been found to be less dependent on dry foods.

Dr. T.J. Dunn D.V.M. "There is ample proof that today's pet dogs and cats do not thrive on cheap, packaged, corn-based pet foods. Dogs and cats are primarily meat eaters; to fill them up with grain-based processed dry foods that barely meet minimum daily nutrient requirements has proven to be a mistake." This parasite is a source of great anxiety among dog caretakers. Thanks in large part to the scare tactics of many veterinarians in promoting preventive drugs, many people believe that contracting heartworms is the equivalent of a death sentence for their dogs. This is not true.

I practiced for seven years in the Santa Cruz, California area, and treated many dogs with heartworms. The only dogs that developed symptoms of heart failure were those that were being vaccinated yearly, eating commercial dog food, and getting suppressive drug treatment for other symptoms, such as skin problems. My treatment, at that time, consisted of switching to a natural (that is, homemade) diet, stopping drug treatment whenever possible, and eliminating any chemical exposure, such as flea and tick poisons. I would usually prescribe hawthorn tincture as well. None of these dogs ever developed any symptoms of heart failure. (danger of: vaccines, commercial petfood, and antibiotics) I concluded from this that it was not the heartworms that caused disease, but the other factors that damaged the dogs' health to the point that they could no longer compensate for an otherwise tolerable parasite load. It is not really that different from the common intestinal roundworms, in that most dogs do not show any symptoms. Only a dog whose health is compromised is unable to tolerate a few worms. Furthermore, a truly healthy dog would not be susceptible to either type of worm in the first place.

It seems to me that the real problem is that allopathic attitudes have instilled in many of us a fear of disease, fear of pathogens and parasites, fear of rabies, as if these are evil and malicious entities just waiting to lay waste to a naive and unprotected public. Disease is not caused by viruses or by bacteria or by heartworm-bearing mosquitoes. Disease comes from within, and one aspect of disease can be the susceptibility to various pathogens. So the best thing to do is to address those susceptibilities on the deepest possible level, so that the pathogens will no longer be a threat. Most importantly, don't buy into the fear.

That having been said, there are practical considerations of risk versus benefit in considering heartworm prevention. The risk of a dog contracting heartworms is directly related to geographic location. In heavily infested areas the risk is higher, and the prospect of using a preventive drug more justifiable. Whatever you choose to do, a yearly blood test for heartworm microfilaria is important.

There are basically three choices with regard to heartworm prevention: drugs, nosodes, or nothing. There are currently a variety of heartworm preventive drugs, most of which are given monthly. I don't like any of them due to their toxicity, the frequency of side effects, and their tendency to antidote homeopathic remedies. Incidentally, the once-a-month preventives should be given only every 6 weeks.

The next option is the heartworm nosode. It has the advantage of at least not being a toxic drug. It has been in use it for over 10 years now, and I am reasonably confident that it is effective. It is certainly very safe. The biggest problem with the nosode is integrating it with homeopathic treatment. But at least it's less of a problem than with the drugs. The last option, and in my opinion the best, is to do nothing. That is to say, do nothing to specifically prevent heartworm, but rather to minimise the chances of infestation by helping your dog to be healthier, and thereby less susceptible. This means avoiding those things that are detrimental to health, feeding a high quality homemade diet, regular exercise, a healthy emotional environment, and, most of all, constitutional homeopathic treatment. Of course, this will not guarantee that your dog will not get heartworms, but, under these conditions, even the worst-case scenario isn't so terrible. If your dog were to get heartworms, s/he shouldn't develop any symptoms as a result. For what it's worth, I never gave my dog any type of heartworm preventive, even when we lived in the Santa Cruz area where heartworms were very prevalent. I tested him yearly, and he never had a problem.

Dogs love Dr. Miller's Detox Tea too! Dogs enjoy drinking Dr. Miller's Detox Tea because it helps improve their digestion, thus eliminate bad breath and body odors. The tea also helps flush toxins and assists in cleansing waste and compacted fecal matter from the colon learn more.

An animal with a healthy immune system will be less likely to become infected with dog worms, including heartworm. Mosquitoes are less likely to bite healthy dogs. In addition, the healthy animals own defence system is able to kill off the larvae of any heartworm that may enter the bloodstream, thus preventing them from reaching maturity and causing harm. Factors that will weaken your dog's immune system include frequent vaccination, commercial pet foods, incorrect diet, stress and even conventional heartworm and other synthetic medication.

Fish Oil Helps Dogs with Lymphoma Live Longer

A diet supplemented with fish oil and the amino acid arginine appears to increase survival time in dogs with lymphoma, a cancer that affects white blood cells. Dogs with this kind of cancer, similar to non-Hodgkin's lymphoma in humans, are easily treated, but as with humans, their cancer tends to return.

Half of the dogs received a special chow with the two supplements in it, and the other half ate chow with soybean oil added. The two chows were identical in nutritional value, and formulated to be equally tasty to the dogs. All the dogs were being treated with the anti-cancer drug doxorubicin every three weeks, and were living at home with their owners.

Previous research has shown that some polyunsaturated fatty acids, like those found in fish oil, may help prevent the growth and spread of cancer tumours, and may help prevent cachexia — the devastating weight loss and muscle wasting seen in some cancer patients despite adequate nutrition. Likewise, arginine supplements have been reported to improve immune responses, and might help the body fight cancer.

The dogs were fed one of the chows twice a day during and after their cancer treatment. The researchers report that compared to the control dogs, those who ate the supplemented chow showed higher blood levels of two fatty acids called C20:5 and C22:6 that seem particularly effective in fighting cancer. Dogs with more of these fatty acids in their blood also tended to have more normal levels of lactic acid, which tends to accumulate in the blood when metabolism is disrupted in cancer patients. The dogs with higher levels of these two fatty acids survived longer than those with lower levels, and had longer remissions, periods of time before their disease came back.

Did you know that the root cause of serious chronic diseases such as heart disease, Alzheimer's disease, cancer, arthritis, and asthma has been identified as chronic inflammation? Although numerous studies have confirmed these findings, few physcians are aware of or consider the fact that the battle against inflammation is at the forefront of the fight for health and well-being of the global population. Authors Joseph Maroon and Jeffrey Bost have set out to reverse that trend with "Fish Oil: The Natural Anti-Inflammatory." For years dog health supplies have included toxic materials to artificially maintain dog health as a temporary measure. The time has come to change to a more natural approach toward dog health. Natural remedies have been shown to achieve more immediate, more complete and longer lasting health, vitality, quality and longevity of dogs lives.

Hypertrophic Osteodystrophy (HOD)

Hypertrophic osteodystrophy causes lameness and extreme pain in young growing dogs, usually of a large breed. Great danes, German shepherds, dobermans, retrievers and weimaraners are examples of breeds that may be affected by this condition. It appears to occur in weimaraners as a vaccine reaction and this may also affect mastiffs and great Danes. In this case, it usually occurs a few days after vaccination and may appear to be worse than the "average" case on radiographs.

HOD usually shows up as an acute lameness, often seeming to affect all four legs simultaneously. Affected dogs may stand in a "hunched up" stance or refuse to stand up at all. They may have a fever but this is not consistently present. They usually have painful swellings around the lower joints on the legs. Some puppies will die from this disease, some suffer permanent disability but many recover later. The disease is so painful that many owners elect to euthanize the puppy rather than watch it suffer, despite the reasonably good chance for recovery, long term. Affected dogs may be so ill that they refuse to eat. If your dog has been on certain antibiotics 2-12 days before your puppy exhibited HOD like symptoms, then it is possible the dog is having an antibiotic reaction and not HOD.

Correctly prescribed homeopathic remedies can often undue the damage caused by vaccines Homeopathy is noted for its success to antidote or remove the toxic effects of vaccines and to re-establish balance in the organism and restore health. Certain homeopathic remedies can minimise vaccine damage. A professional homeopathic vet should be consulted for more information.

Causes, Predisposing Factors and Treatment

Dee Blanco, D.V.M - "You take healthy animals and often very quickly after you vaccinate, you can see simple things like itching of the skin or excessive licking of the paws, sometimes even with no eruptions. We see a lot of epilepsy/seizure, often after a rabies vaccination. Or dogs or cats can become aggressive for several days. Frequently, you'll see urinary tract infections in cats, often within three months after their annual vaccination. If you step back, open your mind and heart, you'll start to see patterns of illness post-vaccination." Health Hazard of Routine Vaccination: placing our animals at risk

Some dogs are sensitive to chemicals and if you feed and water them from a plastic bowl, they ingest some of the chemicals and it can help induce the seizures. There has been reports of a reduction of seizures when the immune system was modulated and the inflammation reduced with Transfer Factor.

Magnesium Deficiency and Your Animal's Health

Holistic veterinarian Roger DeHaan, DVM states that some forms of epilepsy respond to supplementation of vitamin B6, magnesium, and manganese. Learn more about magnesium deficiency in people and animals.

William Pollak D.V.M. - "The most common direct cause of seizures seen in clinical practice in our pets is parasitic infection combined with nutritional deficiencies based on 100% commercial pet food feeding." Animal Epilepsy/Seizure - Causes, Predisposing Factors and Treatment.

"Feeding a natural raw food diet is vital to not only maintaining the health of your pet, but also keeping ideal immune function alive and well. Many times after eliminating seizures through improving the diet, seizures return after commercial pet food is re-instituted. "

"Today's modern approach to dealing with these problems is the administration of more chemicals, injectable or otherwise and even greater processed "prescription" diets. Seizures are masked by giving chemicals that profoundly dull the CNS, slowing it down and confusing it so as to reduce the likelihood of another seizure. These chemicals oftentimes do not work and further confuse the biological system as already described earlier. The underlying imbalance is not directly addressed. Deranged metabolic disorders due to chemical shortages or imbalances are superficially addressed by further limitations in the diet; i.e.even more severely processed foods."

Here is another network that provides additional information on canine epilepsy and other diseases that cause seizures in dogs including canine hypothyroidism. Their canine epilepsy section provides information about canine epilepsy, what happens when your dog has a seizure, possible causes of seizures by age, what tests are used to diagnose canine epilepsy, and information from our Guardian Angels on what they would do differently "if they knew then what they know now." Under medications you will find information a number of medications that are used to control seizures in dogs. Those medications include the more commonly used Phenobarbital and Potassium Bromide as well as newer drugs such as Gabapentin and Felbamate. The section on thyroid contains several articles on canine hypothyroidism and the connection between low thyroid and seizures.

Fulvic Acid Nature's Poison Detoxifier

Fulvic acid acts as an important protective agent. An important aspect of humic substances is related to their sorptive interaction with environmental chemicals, either before or after they reach concentrations, toxic to living organisms. 33 The toxic herbicide know as Paraquat is rapidly detoxified by humic substances (fulvic acid). 34 Fulvic acids have a special function with respect to the demise of organic compounds applied to soil as pesticides. As the most powerful, natural electrolyte known, fulvic acid restores electrical balance to damaged cells, neutralises toxins and can eliminate food poisoning within minutes. When it encounters free radicals with unpaired positive or negative electrons, it supplies an equal and opposite charge to neutralise the free radical. Fulvic acid acts as a refiner and transporter of organic materials and cell nutrients. According to A. Szalay, fulvic acid has the ability to dramatically detoxify herbicides, pesticides, and other poisons that it interacts with – this includes many radioactive elements. This detoxification process may extend to animals and humans, since we are the end-users of these plants.

"Over the years, I began to pull this herb and that herb out of the book, such as the wonderful use of RAW CUCUMBER JUICE topically for CONJUNCTIVITIS, the use of GARLIC internally and externally for ALL TYPES OF PARASITES. I began to use herbal immune tonics, heart tonics, liver cleansers and arthritis herbs. Each year I picked herbs like fruit from a great bearing fruit tree. Whatever I wanted the tree seemed to have for me. All my instincts told me I had stumbled upon the Great Cornucopia of Cure."

The Pet Parasite Cleanse

Pets have many of the same parasites that we get, including Ascaris (common roundworm), hookworm, Trichinella, Strongyloides, heartworm and a variety of tapeworms. Every pet living in your home should be deparasitized (cleared of parasites) and maintained on a parasite program. Monthly trips to your vet are not sufficient. You may not need to get rid of your pet to keep yourself free of parasites. But if you are ill it is best to board it with a friend until you are better. Your pet is part of your family and should be kept as sweet and clean and healthy as yourself This is not difficult to achieve.

Glyconutrients/Saccharides in Veterinary Medicine

Arthur Young DVM, a Certified Homeopathic Veterinarian, uses glyconutrients in his practice on all animals from birds to horses. He always advises his clients to add glyconutrients to their pets diet. Decades of consuming nutritionally questionable commercial pet food products, the proven harmful effects of over vaccination, combined with the unfortunate dependency on steroids and antibiotics have, according to many reliable sources greatly contributed to a national decline in animal health.

The innate ability of the body to heal itself has been seriously compromised in many animals, particularly dogs, cats, and horses. Glyconutrients are compatible with any type of treatment prescribed by both conventional and alternative practitioners. These essential carbohydrates strengthen the body's inner environment which, in turn, enhances the energy that stimulates the immune system, as well as a process called "cell communication". This communication network results in all of the organs working with one another in an orderly sequence of body reactions. Without these activities, good health is impossible to attain. Although the formulation is produced for human consumption, holistic veterinarians have also recognised the importance of glyconutrients in supporting the handling of health challenges in animals. Credit goes to Günter Blobel, M.D., Ph.D. (Rockefeller

University), winner of the 1999 Nobel Prize for Medicine "for the discovery that (glyco) proteins have intrinsic signals that govern their transport and localisation in the cell."

Guinea Pigs, Hamsters, Rabbits Holistic Health and Nutrition

Many rabbit owners and lovers are not aware that natural care and holistic health treatments are available for their pet, and that it can be extremely effective in a variety of conditions. Rabbits have a rapid metabolism and respond well to treatment modalities such as acupuncture, homeopathy and flower remedies. The other benefit of using such therapies is that there are very few, if any side effects, unlike a lot of drugs such as antibiotics, which rabbits tend to be very sensitive to.

Ferrets Natural Health Care and Optimum Diet

To maintain optimum health, ferrets require a diet which most closely resembles that which they would get in the wild. They also require some sunlight. Ferrets are strict carnivores, meaning they are designed to eat whole prey items, which includes all parts of the killed animal. Because of the short GI tract and the poor absorption of nutrients, ferrets require a diet that is highly concentrated with FAT as the main source of calories (energy) and highly digestible MEAT-BASED PROTEIN. The bottom line is that ferrets use fat for energy not carbohydrates and they need a highly digestible meat-based protein not vegetable protein. The most appropriate diet for a ferret would be whole prey foods such as rats, mice or chicks. The worst examples of processed diets are the ferret treat foods

Domesticated Animals (including ferrets and rodents) Suffer from Minerals and Vitamin Deficiency

Our soil, plants, and especially commercial foods are woefully deficient in key nutrients. Scientists theorise that mineral deficiency subjects us, and our animals, to more diseases, aging, sickness and destruction of our physical well-being than any other factor in personal health. Domesticated ferrets, like our domesticated dogs and cats, are prone to suffer from minerals and trace mineral deficiency which makes them prone to diseases. A good source of naturally occuring trace minerals and vitamins for ferrets and other pets are:

Marine Phytoplankton

Dr. Jerry Tennant, M.D. says that marine phytoplankton contains almost everything one needs to sustain life and to restore health by providing the raw materials to make new cells that function normally.

Marine phytoplankton has been called "the most nutritionally dense foods on the planet". Containing a wide range of trace elements, amino acids, vitamins, minerals, chlorophyll, enzymes and cellular materials, marine phytoplankton promotes and maintains optimum health by boosting and supporting all systems within the body.

Seaweed (kelp). Sea Vegetables (Spirulina - Kelp - Chlorella) have been acknowledged as a detoxifyer, a balanced nourishment and a miraculous healing plant. Ocean / Sea algae are the richest natural source of minerals, trace minerals and rear earth elements. Fulvic minerals consist of an immense arsenal and array of naturally occuring powerful phytochemicals, biochemicals, supercharged antioxidants, free-radical scavengers, super oxide dismutases, nutrients, enzymes, hormones, amino acids, antibiotics, antivirals, and antifungals

Bee Pollen contains all the essential components of life. The percentage of rejuvenating elements in bee pollen remarkably exceeds those present in brewer's yeast and wheat germ. Bee pollen corrects the deficient or unbalanced nutrition, common in the customs of our present-day civilization of consuming incomplete foods.

Flaxoil: holistic veterinarians are beginning to recommend to their clients that they supplement their animals diet with a daily dose of flaxseed oil and other nutrients for optimum health and vitality. The vets are finding remarkable results in clearing up skin conditions, relieving arthritic and inflammatory pain, as well as improved over all pet health. (Suzie Zeeman gives flaxoil to her ferrets. She says that their fur looks shinny and thick, and feels soft and smooth since on flaxoil.)"

Bentonite and Pascalite clay: It has helped cows with scours and pneumonia. Veterinarians use it on dogs, cats, horses, etc... for various afflictions including injuries and infections. Pets are helped, too. Transfer Factor Plus - Canine and Feline Formulas are advanced immune support supplements formulated or dogs and cats (also good for ferrets). According to veterinarians, the powerful and proprietary blend of ingredients work together to activate and enhance the immune system's ability to respond to the many pathogens may pets come into contact with. Barley Dog is the original "green" supplement made from the juice of organic Barley grass with a hickory smoked flavor. It provides active enzymes, vitamins C and E, beta-carotene, amino acids, chlorophyll, proteins and essential trace minerals. Regular use can promote healthy skin and coat, reduce bad breath, improve digestion and restore energy levels.

Cindy Engles, Ph.D., documents that for millennia humans have observed animals in the wild eating plants and minerals and applying naturally occurring topical antitoxins from the same sources to combat infectious wounds, parasites and internal disorders. Herds of elephants risk injury and death in a perilous journey to hidden salt caves where they supplement their sodium deficient diet. Monkeys rub poisonous millipedes on their fur to repel biting, disease-carrying insects. Birds line their nests with parasite-resistant herbs. Engel details a world where nature is the pharmacy and every animal is its own practitioner.

On page 83 of Wild Health Cindy Engles, Ph.D. also explains why animals eat their own feces. "Eating feces (coprophagy) is one way animals such as gorillas, elephants, rabbits, and hares add to their supply of essential bacteria through adult life. It is therefore an important aspect of their health care. Overzealous rabbit owners could inadvertently cause disease in their pets by cleaning out the droppings too quickly. By reingesting their own soft pellets (and discarting the hard, dry ones), rabbits extract further nutrients from their food and obtain essential vitamins made in the gut by microorganism."

A chimp with a stomach ache may seek out a broad-spectrum antidote like the bitter-leaf. Or it may swallow whole leaves from several different kinds of hairy plants: leaves with 'Velcro' hooks help clear out intestinal worms. Bears before hibernation and snow geese before migration also turn to Velcro plants. Domestic dogs and cats chew grass, which has the same scouring effect. Many wild animals, and some people, develop 'pica' when ill, a craving to eat earth - particularly clay, which assuages diarrhoea and binds to many plant poisons. Among the most famous clay-eaters are the parrots of the Amazon.

Scarlet macaws, blue and gold macaws, and hosts of smaller birds perch together in their hundreds to excavate the best clay layer along a riverbank. Parrots' regular diet is tree seeds, which the trees defend with toxic chemicals, Canine autoimmune haemolytic anaemia is an autoimmune disease in dogs where the red blood cells are damaged or destroyed by the animal's own antibodies, resulting in anaemia. Haemolysis may take place within the blood vessels or in the spleen. In addition to haemolysis, permanent or intermittent inactivity of the bone marrow is common. According to information contained on the leaflet accompanying the Rabdomun rabies vaccine, extraneous proteins found in some vaccines can cause autoimmune disease.

Aromatherapy

Aromatherapy is the art and science of using from plant sources for health and wellbeing. Essential oils have a potentially powerful healing capacity. The early civilizations — Egyptian, Greek, Roman, Chinese, Egyptian, — made use of aromatic plant materials in religious ritual and to promote physical and mental wellness. Hilary Jupp, a former breeder of Wolfhounds, is now a judge for the breed at championship show level. Her main interests (besides Irish wolfhounds!) are in the field of nutrition and holistic therapies for animals. She is a bio-energy therapist, a Reiki Master, has a Doctorate in Radionic Medicine, and is especially interested in channelling, animal communication, flower essences, and homoeopathy. She has written the Breed notes for the British weekly paper "*Dog World*" since 1986, has written for it on natural therapies, and for several years wrote for Hoflin Publishing's "*Irish Wolfhound Quarterly*" and "*Windhound*". In her wonderful web site you will find of information about alternative therapies for animals such as:

Acupressure, Bio Energy Therapy, Bowen Technique, Chiropractic, Massage, Osteopathy, Tellington Ttouch, and Animal Links. Hilary also list various natural treatments for various conditions . Acupuncture was first used to treat animals over 3000 years ago in Asia. An elephant had a stomach disorder similar to bloating. The healer used slender needles placed in specific positions on the skin to relieve the pain and to help the elephant. Since that time acupuncture has been used all over the world to alleviate many kinds of health problems for many species of animals.

Temperament and Behavioural Problems, Cancer, Epileptiform fits, Heart disease, Digestive disorders, Skin disorders, Bone and joint disorders and more.

- Preventing and Healing Animal Cancer, arthritis, and other disease with flaxoil
- Amazing Healing Clay: a wonderful nutritional supplement for animals
- Orthomolecular/Nutritional Medicine for Animals
- For Cats and Dogs' Optimum Health: Sprouting and Grasses
- Adding Kelp meal to your pet's diet to promote optimum health
- Ready to Eat Health-Food For Your Pet: commercial prepared, ready-to-eat frozen or freeze dried RAW meals for dogs, cats and ferrets. (if homemade food is not feasable).

"For a return to health, pets require a diet which strengthens the immune system and most closely resembles that which they would get in the wild".

Frustrated with the Failures of Conventional Veterinary Medicine: "After 10 years of traditional veterinary practice I became tired of having no treatment for chronic disease, incurable conditions, and a plethora of allergic maladies which seem to plague all veterinary practices. I was frustrated with giving animals cortisone because I had no other solutions, or using antibiotics for infections which I knew were of viral origin. Amazing animal health improvements and cures of cancer and other chronic disease with natural, alternative health care and optimum nutrition.

This is a list of professional holistic veterinarians and animal consultants who are willing to offer phone consultations. (If there are no holistic veterinarians in your hometown, phone consultations for homeopathic treatment can be just as effective as an office consultation, especially if you already have a diagnosis of your animal's condition.)

This is a group where sharing of information on the application of homeopathy (for people and animals) is given in a practical, usable format. Dedicated to the herbal, natural, alternative, and holistic care and keeping of livestock, animals and pets. Dedicated to the herbal, natural, alternative, and holistic care and keeping of livestock, animals and pets. Great for small farm owners and homesteaders. Although goats are my first love, we raise and chat about all sorts of animals from the farm.

This is a list whereby all who are interested in healing with herbs and natural methods can voice their opinions and share their views. The predominant theme is "Herbal Remedies" and "Natural Healing" for you goats. We welcome the discussion of other animals but the list is primarily directed at alternatives or supplementing for your caprine friends. This is for seeking other ways , i.e.. Alternatives and/or supplementation to conventional medicine for those who would like to seek other methods of naturally raising their stock. It is not restricted to Classical Homeopathy, and may , to a degree, include adjuncts such as diet + nutrition, herbs, lifestyle etc., and health related topics such as vaccination etc.

This list will contain information pertaining to dog health (both in pure breeds and mixed breeds of all ages), primarily with regards to drugs that can or have caused illness and/or death in dogs. The focus is presently on the drug Rimadyl (carprofen). However, all issues regarding dogs' health can be addressed at this site. Only those genuinely concerned and seriously committed to the welfare of dogs are encouraged to join the list.

For Veterinarians and for those with care of pets and animals. For pet owners: Introduction to Veterinary Homeopathy. The Elizabethan collar is a type of plastic collar that is effective in preventing pets from licking or biting themselves. The only time I will use such collar is to prevent my animal from tearing up surgical sutures, until healing is complete. I will never use such collar to prevent my dog or cat from licking skin eruptions, hot spots, or inflammed and infected area on their bodies. The built in germicide found in the saliva of a dog or a cat amazingly heals wounds in record time. Any injuries are carefully licked and coated with antiseptic saliva from the tongue. This aids in preventing infection entering into a wound.

Natural Herbal Digestive Aids for Dogs and Cats

"Most of what I've learned about the holistic care and treatment of my dog I've learned from Shirley's Wellness Cafe. This vast and detailed resource is the result of a journey of natural healing that webmaster Shirley used on herself and her family to completely turn their lives around. Those lessons are applied here with full force for the benefit of our canine companions. Raw food diets, detailed feature articles on a wide variety of wellness topics, health reports, and studies are all represented here in a slightly cluttered, yet home-style approach that blends a first-person knowledge and passion for the subject with a caring and nurturing writing style. Thank you, Shirley. The 'net's a better place because of sites like yours

Animal Husbandry and Cattle Management in *Arthasastra*

Humans had to interact closely with nature for their basic needs since prehistoric times; their daily life was totally dependent on plants and animals. Thus the relationship between humans and animal is very ancient. *Atharvaveda* has several references to dairy farming, cattle health care etc. Asoka, the Buddhist emperor (300BC), established a network of veterinary hospitals throughout India. Kautilyayys *Arthasastra* (321-296 BC) describes a well-managed animal husbandry and cattle management system. In this article I have dealt with ancient animal husbandry from the *Arthasastrayy*s point of view.

We are amazed at the detailed and scientific regulations for seemingly such a minor activity 2300 years back. It also shows the concern of the state not only to exploit the animals but also to live with them symbiotically. Large cattle-sheds were well regulated, and so were the wild game sanctuaries where animals were safe from poaching. Even for animals, a medical ethics was enforced by *Arthasastra*.

Let us first introduce Kautilya and his *Arthasastra*. All sources of Indian traditions -Brahmanical, Buddhist and Jain - agree that Kautilya destroyed the Nanda dynasty and installed Chandragupta Maurya on the throne of Magadha. The name Kautilya denotes that he was of the Kutila gotra. The name yyChanakyayy means that he was the son of Chanaka, though yyVishnuguptayy was his personal name. *Arthasastra* was composed by Kautilya in probably between 321-296 BC, though there are doubts about the date of the composition of *Arthasastra*. A workshop held by the Indian Council for Historical Research, Delhi, concluded that the *Arthasastra* in its present form was a compilation made by a scholar, Kautilya, in 150 AD.

But there is no doubt that Chandragupta Maurya ascended the throne around 321 BC. *Arthasastra* had never been forgotten in India, but the text itself was not available until, dramatically, a full text on palm leaf in the *grantha* script, along with a fragment of an old commentary by Bhattasvamin, came into the hands of Dr. R. Shamasastry of Mysore in 1904. And finally he published the text not only in Hindi but also in English in 1909. *Arthasastra* is a valuable document, which throws light on the state and society of India at c. 300 BC. Here we can say that Kautilyayys *Arthasastra* is totally an administrative text of ancient India, especially of the Mauryan times. Animal husbandry or cattle management was one of the main administrative jobs of the state. Kautilyayys *Arthasastra* also throws light on ancient cattle management practices of the country.

Animal Husbandry

Arthasastra contains an elaborate analysis regarding various aspects of livestock with prescriptions for their better management. Kautilya was also highly conscious about the benefits man gets from different species of animals. Both domestic and wild animals were well protected by all means in Kautilyayys time. Kautilya mentioned a variety of animals such as deer, bison, birds, fish, cattle, elephants, horses, asses, pigs, camels, sheep, goat, etc. According to *Arthasastra,* cattle rearing was the second most important economic activity. Cow and she-buffaloes were reared for milk and the bulls and he-buffaloes were used as drought animals. Ghee, which had the advantage of being easily stored and transported, was the main end product. Cheese was supplied to the army, buttermilk was fed to dogs and pigs and whey mixed with oilcake was used as animal feed. Wool was obtained from sheep and goats.

Cattle Superintendent and Cowherds

Crown herds were the responsibility of the chief superintendent who either employed cowherds, milkers, etc. on wages or gave some herds to a contractor. Chief superintendent was responsible for cattle (cows, bulls and buffaloes), goats, sheep, horses, donkeys, camels and pigs. He had to keep a record of every animal in the different types of herds, of the total of all such animals, of the number that die or are lost, the total collection of milk and ghee and other products. Private herds could also be entrusted to the state for protection on payment. Private owners of animals paid a sales tax of a quarter *pana* for every animal sold. The superintendent of cows had to supervise the maintenance of cows as well as bull, oxen, buffaloes, and their young calves.

The duties of a superintendent were: supervise herds and obtain exact information about abandoned and useless herds; maintain class of herds by registering each animalyys natural and branded marks, colour, and distance from one horn to another; update the number of stray cattle and register total number of cattle lost permanently; collection of information about total production of milk and ghee or clarified butter, etc.

The duties of cowherds were: graze their herds in forest grounds under suitable guards and in groups of ten according to their types; treat the animals during the period of their illness; take care of animals when animals went to river or lakes to drink water; to ensure that the watering place is safe and free of crocodiles and not muddy; to milk timely; inform the owners about the loss of cattle and cause of loss; to hang bells around the necks of timid animals in order to frighten snakes and to locate them easily when grazing; in rainy season, autumn and early winter, cows and she-buffaloes were to be milked twice a day and in late winter, spring and summer only once a day.

Generally four types of herd were recognised in the *Arthasastra*. Those:

- *Looked after by attendants:* - the chief superintendent employed, for each herd of 100 animals, one cowherd or buffalo herdsman, a milker, a churner and a hunter guard who would protect the herds from wild animals.
- *Looked after under contract:* - a balanced herd of 100 animals could be given to some one on contract.
- *Unproductive animals:* - a balanced herd of 100 animals may be given to someone for an annual payment, related to what the herd can produce.
- *Private cattle looked after for a share:* - private cattle owners could place their animals under the protection of the King on payment of a protection share.

Basically cowherds played an important role in the maintenance of livestock. On the other hand, the herdsmen had to pay one-tenth of dairy produce to the superintendent of cows. Therefore, such supervision was a source of royal income.

Official Breeder

Animal breeding was given special attention in Kautilyayys *Arthasastra*. The King appointed an official breeder for improving the breed of animals. According to *Arthasastra*, for breeding purpose, the following proportion of male animals should be kept for every herd of 100 animals:

Accounting of Animals

According to Kautilya, the Chief Superintendent should keep an account of the animals as follows:

- All calves should be identified (by branding or by nicking the ear) within a month or two of birth; all stray cattle be marked if they remain unclaimed for more than two mounts.
- Every animal should be identified in the records with the details of the branding mark, any natural identification marks, the colour, and peculiarity of horns.
- An account should be maintained of cattle lost.

Animal Welfare

There is extensive evidence in the *Arthasastra* about Kautilyayys concern for the welfare of animals. Regulations for the protection of wild life, a long list of punishment for cruelty to animals, rations for animals, rules for grazing and the responsibility of veterinary doctors are some of the major topics.

- Animal sanctuary was the remarkable feature of *Arthasastra*. As part of creation of the infrastructure of a settled and prosperous kingdom, an animal sanctuary, where all animals were welcomed as guests was established.
- Killing or injuring protected species and animals in reserved parks and sanctuaries was prohibited. Even animals, which had turned dangerous, were not to be killed within the sanctuary but had to be caught, taken outside and then killed.
- Sea fish which had strange or unusual characteristics, fresh water fish from lakes, rivers, tanks or canals; game birds or birds for pleasure such as curlew, osprey, swan, pheasant, partridge, parrot and mynah; and all auspicious birds and animals were declared as protected species.
- The chief protector did not allow the catching of fish, or the trapping, hurting or killing of animals whose slaughter was not customary.
- Live birds and deer received as tax were let loose in the sanctuaries.

- If protected animals or those from reserved forests strayed and were found grazing where they should not, they were to be driven away without hurting them. Stray cattle were to be driven off with rope or a whip without harming them.
- Butchers paid tax for sold meat at the rates given below:

Kautilya ordains that Butchers should follow some rules and regulations, like: should sell only freshly killed animals. The sale of swollen meat, rotten meat and meat from naturally dead animals was prohibited. Fish without head or bones should not be sold. Meat may be sold with or without bones. If sold with bones, equivalent compensation (for the weight of the bone) was to be given.

Grazing Places or Pastures

Pastures were basically required for grazing domestic cattle during ancient India when stall-feeding was rarely adopted. Pastures located in the forest were relieved from danger of tigers, beasts, and thieves. A separate clause was prescribed by Kautilya to protect livestock in pastures and defined the penalties for promoting imprudent grazing of pastures. Pasture lands within the village boundary were the responsibility of the village headman. He collected the charge for grazing on common land and ensured that cattle do not graze or stray into cultivated private fields or gardens or eat the grains in stored sheds. Headman had the responsibility for collecting the revenue for the village from the charges levied on grazing in common land, from the prescribed fines and the fines levied by the state.

According to *Arthasastra*, bulls belonging to village temples, stud bulls and cows up to ten days after calving were exempt from payments from the following grazing charges:

Healthcare of Animals

Kautilya set forth guidelines for providing medicine to livestock. Various mixtures were prescribed to cure different diseases. He maintained that cowherds shall apply remedies to calves or aged cows or cows suffering from diseases. Regarding the dosage he mentioned that proportion of a dose is as much as an *aksha* (quantity) to men; twice as much to cows and horses and four times as much to elephants and camels. Kautilya had provided the facility of veterinary doctors, who were to cure the ill animals. But Kautilya fined the veterinary doctors when the condition of sick animals became worse. If the animal died they had to repay the cost of animal. We thus see that the medical ethics was already enforced by the state in ancient India.

Nutrition Management of Cattle

Kautilya had analysed many cattle problems and set the guidelines:

- Drought oxen would be provided with subsistence in proportion to the duration of service rendered.
- Milch cows would be provided with subsistence in proportion to the milk obtained from them.
- All cattle would be provided fodder and water abundantly.

Animal Products

Kautilya even standardised the purity of the animal products:

Ghee: Normal yield of ghee from 1 drona of milk:

Kurchika- Cheese (to be supplied to the army)

Kilata- Whey (mixed with oil cake, used as animal feed)

Butter-removed buttermilk- Not for human consumption, to be fed to dogs and pigs.

Such as hair, skins, bladder, bile, tendons, teeth, hooves and horns were useful products of wild animals form the forest.

Importance of Elephants and Horses

In *Arthasastra*, elephants and horses played a major role in defence practices. According to *Arthasastra,* the best army had best horses and elephants, of good pedigree, strength, youthfulness, vitality, loftiness, speed, mettle, good training, stamina, a lofty mien, obedience, auspicious marks, and good conduct. Kautilya categorised several kinds of elephants and their physical characteristics: war elephants, riding elephants, and untrainable elephants. The state took into account the physical characteristics of elephants with red patches, evenly fleshed, of even sides and rounded girth, with a curved backbone and well endowed with flesh. The following structure of army shows the importance of elephants and horses:

SmallAnimal Diseases

Louise Pasteur, a Frenchman who was neither a physician nor a veterinarian moved into the spotlight to help find a vaccine forRabies. He began the study ofRabies when two rabid dogs were brought into his laboratory. One of the dogs suffered from the dumb form of the disease: his lower jaw hung down, he foamed at the mouth, and his eyes had a rather vacant look. The other dog was furious: he snapped, bit any object held out to him, and let out frightening howls (McCoy 65).

Through the studies already observed,rabies was transmitted through the bite of a rabid animal, and that the incubation period varied from a few days to several months. Beyond this, nothing definite was known. Then M. Bouley, a professor of veterinary science, noted a germ or organism in the saliva of a rabid dog. Pasteur confirmed Bouley's findings by collecting some mucus from a child bitten by a rapid dog, and injecting it into rabbits.

The results of this experiment ended with all the rabbits dying within 36 hours. This experiment established two facts: an organism was present in the saliva of rabid animals, and it could be transmitted to another animal or a human being through a bite (McCoy 66).

Further research led Pasteur to the conclusion that therabies organism was located in other parts of the infected animal's body besides its saliva. Experiments on the skulls of rabid dogs shoed that the brain contained therabies virus. Pasture then cultured some viruses from several rabid dogs' brains. The virus was then injected into rabbits. In every case therabies would appear within 14 days (McCoy 67).

After several experiments, Pasteur went on to perfect arabies vaccine. He first demonstrated to physicians and veterinarians that therabies could be cultured from the brains of living dogs.

Pasteur successfully proved that his antrabies vaccine could now be safely administered and animals could be vaccinated against the disease.

Chapter 2

Animal Diseases and Their Control

Animal disease is a state of discomfort associated with disturbed or abnormal functioning of animal body due to defective heredity (genetic diseases), defective nutrition (deficiency diseases), unsuitable environment (environmental discomforts) and pathogens. Defective heredity or genetic diseases can be overcome only through selective animal breeding. Nutritional disorders are overcome by providing balanced^ diet in optimum amount. Environmental discomforts can be removed by providing suitable shelter. Domesticated animals are attacked by a number of pathogens of various , types â€"viruses, bacteria, protozoa, fungi and worms/parasitic animals. The diseases caused by pathogens that can be transferred from one individual to another are called infectious diseases (communicable diseases).

Infectious or communicable diseases are of two types (/) Contagious Diseases. The diseases spread from diseased animals to susceptible animals through either direct contact or indirect contact with articles/ materials touched by diseased animals, *e.g.,* foot and mouth disease, *(ii)* Non-contagious Diseases. An intermediate agent is required for transmission of pathogen from diseased animal to healthy susceptible animal. It can be vector *(e.g.,* housefly, mosquito, tick), vehicle, feed, water, air, etc. Pathogen enters the body through skin, digestive tract, respiratory tract, urogenital tract, egg, teats, udder, umbilicus, placenta, conjuctiva, etc.

For preventing the spread of infectious diseases, care should be taken about (/) Isolation of infected animals, (ii) Disinfection of the animal house and articles used for lending to the animals, *(iii)* Proper disposal of dead infected animals and all contaminated articles, (iv) Vaccination of animals against all major diseases, (v) Injection of antiserum to all animals against the disease in case of spread of an infection, (v/) Use of alternate grazing grounds in case a pasture was being used by diseased animals. *(vii)* Whenever an infectious disease appears, the veterinary authorities should be informed so that they can promptly initiate measures to check the spread of disease.

Animal Diseases

Both domestic and wild animals carry many diseases, parasites, and viruses; however, the majority of those diseases cause little harm to humans. Three of the most common and serious illnesses that are transmittable to humans are listed below.

Rabies

Rabies occurs when a virus attacks the brain in warm-blooded animals. Once an animal or human contracts the virus and symptoms appear, the victim always dies. Reptiles and birds cannot contract or transmit the rabies virus. Animals with rabies can survive for months before becoming sick and dying. During that time, the infected animal can pass the virus onto other animals and humans. You do not have to be bitten to contract rabies. The virus can be spread through any open wound or through direct contact with the eyes. By far, the rabies virus is one of the biggest concerns in our area. Raccoons, fox, and bats are the usual carriers, however, all warm-blooded animals can contract and transmit the disease. Yearly, hundreds of residents in our area receive rabies vaccinations after being exposed to the virus by a wild animal or their own pets.

Occasionally domestic dogs contract rabies, however, one of the biggest risks to residents are free roaming cats. Residents who do not immunise their dogs, cats, and ferrets and who allow their cats to roam freely are placing themselves and their families at risk for rabies infection. Residents in our area should educate themselves about the rabies virus and learn how to protect their families and their pets.

Rabies Clinics are held regularly at the at a cost of $10 per domestic animal. Vaccines are not administered to pets that are considered wild animals. There is no proven rabies inoculation for the prevention of rabies in wild animals. Falls Church City residents are welcome at the clinic. Contact the at 703-931-9241 to confirm the clinic is being held.

Reptile-Associated Salmonella

Salmonella is a natural bacterium found in the gut of reptiles and snakes. Unfortunately, when the animal sheds the virus in their feces, their owners can contract it simply by handling the animal. It is illegal to own reptiles in Falls Church City. Owners of reptiles, lizards (especially iguanas), and snakes should familiarize themselves with this condition.

Psittacosis

Also known as "Parrot Fever," psittacosis is spread by a bacterial infection in birds and can be spread to humans. Psittacosis is acquired by inhaling dried secretions from infected birds. The incubation period is 6 to 19 days. Although all birds are susceptible, pet birds (parrots, parakeets, macaws, and cockatiels) and poultry (turkeys and ducks) are most frequently involved in transmission to humans.

Once infected, symptoms found in humans include fever, chills, headache, muscle aches, and a dry cough. Humans can also contract severe pneumonia and other health problems. Psittacosis has been found in pet parrots in Falls Church City in the past.

To Keep Your Pets Safe:

- Innoculate your pet dogs, cats, and ferrets for rabies. Do not let your pets roam free as they are more likely to encounter a sick animal without your knowledge. Be especially careful when letting your pets out at night.
- Do not feed wildlife. Observe wild animals from afar and never approach or try to pet a wild animal. Do not keep wild animals as pets. It is illegal and dangerous to expose your family to these animals. Educate your children about the dangers of approaching or petting wild animals.
- If you should see a wild animal that appears to be overly friendly, staggering, with a wobbly gait, or just appears to be sick, get away from the animal and contact the Animal Control Officer or the Police Department immediately.
- If you are bitten or scratched by any wild animal, contact the Animal Control Officer or Police Department immediately. Wash the wound immediately with soap and water. If possible, try and see where the suspect animal goes after the altercation. Rabies is fatal and can incubate in humans for months. If bitten or scratched, SEEK MEDICAL HELP IMMEDIATELY!
- If your pet is in an altercation with a suspected rabid animal or any animal, minimise your contact with your pet until a veterinarian can examine the animal.

Animal Disease Control Issues, Options, and Impacts

Animal disease poses an economic threat to livestock producers that has few parallels in the world of crop farming in terms of the speed of onset, the potential severity of outcomes, the rewards for success, and the need for regional as opposed to farm-specific solutions. Animal diseases result in increased mortality and morbidity in livestock populations. Disease may affect performance through reduced fertility, delays in reaching maturity for reproduction or sale, decreased production of milk, eggs, or wool, decreased draught power, or decreased weight of fattening or cull animals. Animals raised by small-scale producers and backyard farmers in developing countries tend to be plagued with reinfection, and they typically lack access to diagnosis and control programs. Diseases are also easily spread across farms, regions, and national boundaries. The proximity of small producers to larger-scale operations, despite the latter's best efforts at control, creates a potential for constant reinfection. As noted in the Thai country study (Poapongsakorn *et.al.*, 2003), backyard livestock producers often do not strictly follow recommended procedures and enforcement is difficult in a small-farm setting.

Epidemics affect not only farmers, but also the entire agricultural sector and even the national economy, as recent experience in the United Kingdom has shown. Consequently many countries have implemented eradication and control programs to combat epidemics and prevent the introduction or reintroduction of contagious diseases, supported research and development on vaccines and, in some cases, produced vaccines to farmers. These programs would be justified even in closed economies, but take on added luster if disease control increases the access of poor countries to developed country markets.

Recognising that animal health is truly a global issue, animal health officials worldwide coordinate their animal disease control strategies with the *Office Internationale des Epizooties* (OIE), also known as the World Animal Health Organisation. OIE is the international body in charge of monitoring animal health and minimising the spread of important animal diseases that affect productivity. Through this organisation, an International Animal Health Code has been developed that lists various disease statuses within every country and the time frame and conditions for a country or zone to be considered disease-free under each category. Besides providing better animal health status domestically, compliance with OIE norms can result in a certification of compliance with developed country standards, opening a whole slew of new export possibilities. Among the countries studied, Brazil and Thailand are at the forefront of this high-value trade in livestock products.

There are two major classes of disease recognised by the OIE. List A diseases are defined as being transmissible diseases that have the potential for very serious and rapid spread, irrespective of national borders, are of serious socio-economic or public health consequence, and are of major importance in the international trade of animals and animal products. Horst *et al.,* (1999) report that of the outbreaks that occurred in 1997, the most frequently reported List A diseases were Newcastle disease, Foot-and Mouth Disease (FMD), and Classical Swine Fever (CSF).

Fifty-nine countries reported one or more outbreaks of FMD, 38 countries reported one or more outbreak of CSF, and 87 countries reported one or more outbreaks of Newcastle disease. List B diseases are transmissible diseases that are considered to be of socio-economic and/or public health importance within countries and that are significant in the international trade of animals and animal products. Many of the list B disease are also considered a significant public health concern because of their zoonotic nature. Not all of these are feed or food-borne. For instance, trypanosomes is a major zoonotic concern, but is not feed or food-borne.

Most countries monitor the movement of animals and animal products, which includes veterinary checks at borders. To prevent the spread of disease within a country or across borders, rejection of animals and animal products may occur at port of entry (or at quarantine station) if the quarantined animal or commodity originates from a country or region from which the imported product is prohibited, an OIE List A or B disease is found, or if the quarantined commodity is found to be unhealthy or in poor condition. Similarly, animals will be destroyed if they fail subsequent treatment.

Historically trade decisions on animal products were based on the country status for a disease. Countries that had List A or B diseases anywhere within their border were not allowed to trade either live or animal products from that animal (or family of animals). Over time it has been recognised that in many countries, there are identifiable and measurable gradations in the degree of risk presented by imported animals and animal products, and that these gradations are more often tied to climatological, geographical, and biological factors more specific to eco-regions than to national political boundaries. This provided the basis for certifying parts of countries disease-free for trade purposes, which only began in 1995 with the adoption of the Sanitary and Phyto-Sanitary (SPS) agreement in the wake of the establishmentof the World Trade Organisation (WTO) in late 1994. The adoption of regional certification enables trade from a specific region within a country that may be free of a specific disease, even if the rest of the country still is considered as having the disease. Regional certification has been attempted in most developing countries wishing to establish livestock export zones. Establishment of such a disease-free zone greatly increases pressures on small-scale producers within the zone to make investments in disease control more typically absorbed on larger farms.

Options for Disease Control

Though many diseases can be controlled through good hygiene, many animal diseases are controlled through either vaccination or a stamping-out policy, which includes depopulation (slaughter and disposal of potentially infectious material), quarantine of remaining stock, and continued surveillance of areas where disease had been observed. The method by which a country chooses to combat a disease may have impacts on small-scale producers if either the full cost of the control program is directly passed onto them with no subsidy from the government, and/or the political environment is such that they "disappear" as they are "forced" out of production, in an effort to eliminate disease in a region desiring to obtain regionally free disease status.

Vaccination programs as preventive measures are among the most often practiced methods for controlling disease, particularly for high-risk countries. Problems arise when a country lacks the infrastructure to make and deliver vaccines. Countries with large-scale operations usually rely on the established private sector to deliver vaccines; often the public sector is the main source of vaccines for small-scale and backyard producers. An indirect cost associated with vaccination policies is that disease-free (without vaccination) countries are reluctant to import livestock and livestock products from countries using vaccinations. This is because vaccinated animals can potentially transfer a virus used to create a vaccine to the importing countries own livestock population. Thus, the choice of method for disease control in more limited in exporting countries, as the form of control chosen will affect market access.

Stamping-out programs involve the eradication of a disease by the destruction of all infected animals. When outbreaks occur, protection and surveillance zones are established around the outbreak, and animals in the protected zone are destroyed. The level of surveillance in the area is increased and the movement of animals from surveillance zone is restricted. Stamping-out programs tend to be applied when there is no known vaccine or it is not available, when the disease has reached a low level of incidence following other forms of control, or when a country wants to maintain its access to export markets that require certification of being disease-free without vaccination.

If there is no compensation for stamping-out, then producers, particularly small-scale producers, are reluctant to participate and if they participate it may mean that they no longer can afford to produce. In order to avoid de-capitalisation, small-scale producers who rely solely on their animals for income may move their animals across the border rather than killing them, further spreading infection. If the large-scale operations have enough political clout, they may use the state to force small-scale producers out of business, or alternatively may choose to work with them in eradication. The latter has occurred in Mexico, where large-scale producers in Sonora and the Yucatan aided the small-scale producers in eradicating disease in order to obtain regionally-free disease status with regard to hog cholera (Narrod, 1996).

The method used to combat a disease outbreak depends on the severity of the outbreak, the degree to which total eradication as been achieved, and the number of animals that are slaughtered. Different methods affect participants to a different degree. Table 3.6 summarises some of the economic impacts of the different control programs chosen and who bears the cost under different scenarios. It can be seen that the incidence on different parties depends on the details of the method chosen. It is also clear that-as in the case of the environment-major negative externalities are present in the area of animal health and disease control.

Vaccination alone does not achieve eradication of a disease, as everyone must comply for the program to be effective and the program must be maintained over time, and disease inflows from other areas prevented. Since producers cannot recoup the full social benefits of investments in disease control, an externality is created (Umali, Feder, and deHaan,1992; Hirshorn, Unnevehr, and Narrod, 2000). An individual producer may take steps to eliminate an animal health problem, but cannot do so completely without the cooperation of other producers. On the other hand, producers can benefit for m the control efforts of others without contributing to their cost. This creates a public good problem and the under-production of animal health efforts, and may warrant public sector intervention.

As with environmental externalities, governments should in principle intervene in the market to correct for the animal health risk associated with the production of goods that create a negative externality. Without intervention, society bears a cost that may not fully be reflected in the local market price for the agricultural products and inputs used in producing them. The government may instead choose to intervene and vaccinate the whole affected population so that there will be minimal time wasted in reaching disease-free-status to enable access to trade. This is especially valid when there are producers that cannot or refuse to participate otherwise.

Many countries have regulatory agencies in place to remedy the failure of the free market to allocate resources efficiently in the area of animal health. These agencies are supposed to intervene to reduce the social costs associated with animal health problems. But public intervention is costly, enforcement is difficult, and once established, public systems can fail to respond to changing market conditions. This can be aggravated by lack of strong institutions and political clout to monitor disease status effectively. Difficulties in eradication of disease may also be exacerbated with many small-scale and backyard producers, infected wildlife, smuggling, and cockfighting. In such situations private solutions to animal health problems can emerge, where the costs of information or monitoring can be overcome.

A key scale difference in animal health externalities is that the rationale for public intervention is stronger where a given concentration of animals per square km are kept by smallholders as opposed to a few large farms. Large-farms will find it easier to match contributions to a voluntary association better to benefits achieved than would be the case for a multiplicity of small farms, where it is easier to be a free rider.

Animal Disease Status and Scaling-up in the Case Study Countries

Among the study countries, certain OIE List A diseases are of particular concern from the standpoint of production and trade: FMD, CSF, Rinderpest, and Newcastle disease. The study countries are in the process of eradicating these diseases and some have been more successful than others. Rinderpest has been eradicated according to OIE in both Brazil and the Philippines. Eradication efforts in Thailand have been successful and the Thailand's Rinderpest free status confirmed in 2001 by the OIE, but Thailand is required still to reconfirm this status annually to the OIE. Since March 1998, India has declared itself provisionally free from Rinderpest; but OIE to date has not officially confirmed disease-free status. To achieve the final stage of freedom from the disease, the Indian government has initiated a National Project on Rinderpest Eradication.

Foot-and-mouth disease is one of the most contagious diseases of cloven-hoofed animals.. All four-study countries have actively been trying to eradicate FMD from affected regions. Of the study countries, the Philippines presently has FMD-free zones in Mindanao, Visayas, Palawan and Masbate, which have been approved by the OIE and where vaccination is not practiced.

Brazil is currently recognised by the OIE as FMD-free with vaccination in the states of Bahia, Espírito Santo, Goiás, Mato Grosso, Mato Grosso do Sul, Minas Gerais, Paraná, Rio de Janeiro, Rio Grande do Sul, Santa Catarina, São Paulo, Sergipe, Tocantins, the Federal District, and Rondonia. However, FMD was reported in August 2000 in Rio Grande do Sul, and suspected to have come from infected animals from across the border in Paraguay, which has delayed their status of becoming free without vaccination. It shows the latest incidences of sanitary problems in Brazil. Note that these areas are where most of the livestock production is currently located. The northern part of Brazil has difficulty in eradicating disease due to its proximity to the Amazonian forest region. The other disease concerns in Brazil where there are sporadic outbreaks are vesicular stomatitis, CSF, and Newcastle disease.

There has been a systematic effort to eradicate FMD from Thailand. Currently the Thai government is promoting 3 regions (Regions 2, 8 and 9) as disease-free zones and have set a large number of quarantine stations to prevent the spread from within the country and especially from outside the country. Region 2 is the export-oriented zone and it is where the largest operations are currently found. However, FMD outbreaks continue to emerge sporadically in the country and OIE has not recognised these regions as disease-free.

In case of swine, Kehren, Murphy, and Tisdel (1997) claim that several factors have seriously affected the spread of disease among swine, particularly at village level. These factors include lack of capacity for accurate diagnosis at the village level, no restriction in movement of sick animals, the high density of swine populations, unhygienic conditions and poor sanitation, etc.

In Thailand, the other main animal health disease concern is CSF. Classical Swine Fever outbreaks appear to be mainly a problem in Region 7, where there is the highest pig density. Though Newcastle disease has been a problem in the past due to a large number of backyard producers, with contract farming and the hygienic controls integrated firms require, the last case was reported in 1996.

FMD is a major disease facing Indian livestock and it is prevalent all over the country. Part of the problem is that there has been no systematic control and vaccination program against FMD in the country, even though there is a massive but sporadic vaccination program. The population at risk in the country (all susceptible species) is about 420 million, and barely 5 percent of the animals at risk are vaccinated.

Besides the usual funding problems of public-sector initiatives in developing countries, efforts to date have suffered from failure to offer compensation to affected farmers whose stock is destroyed. To compensate for this, the private sector in the Brazilian states of Santa Catarina, Mato Grosso do Sul, and Parana, has set up insurance- like programs in partnership with the government. In the event of quarantine or the need for slaughter of diseased animals, the producer is compensated by a fund, which producers pay into in better times, and is supplemented by the government. In Thailand companies such as the Betagro Group, the second largest company in the chicken and swine farming business, has established a laboratory to work with the producers to find out the causes and effects of diseases.

The case studies report mixed evidence of disease control and scaling up production. All countries appear to be currently having difficulties in with disease eradication regardless of size of producers in cases (particularly in Brazil and Thailand), where there is smuggling of animals across borders, making it difficult to eradicate diseases such as FMD and CSF. Conversely the Philippines, being an archipelago, has had more success in being able to eradicate disease regardless of farm size.

The Thai study (Poapongsakorn *et. al.*, 2003) alludes to small-scale farmers being less interested in eradicating non-stock-life threatening diseases (such as FMD) than life threatening disease such as Newcastle Disease, making it difficult to totally eradicate FMD from the country. Cases where the success in disease eradication has been greatest is where the main gainers from disease eradication are the large integrators who are exporting. However, whenever export is banned, small-scale farmers who serve as subcontractors suffer the most since the large integrators tend to react to such a ban by reducing the amount of contract broiler production before reducing their own production.

In sum, animal disease control issues are growing with increased concentration of animals in restricted space around major cities and other producing areas. This is a problem for both large and small producers, but it is harder to monitor compliance with disease control measures where the production system is primarily smallholder. It is also difficult to separate small-scale and large farms for the purposes of disease eradication unless then are physically separated by zone. The rise of disease-free export zones, particularly those certified for export without vaccination, increases the incentive of large-scale export producers to either eliminate small-scale producers from the zone or co-opt them into a system of disease control that is similar in terms of overhead costs to those practiced on much larger farms, resulting in relatively high per unit costs of production for smallholders in those zones, and possible competitive disadvantage.

Animal Welfare and Effect of Changes in Study Countries

Increasingly animal welfare issues have been raised by consumers in developed countries-especially in the European Union (EU). The two meat-exporting countries of the four country sample-Brazil and Thailand-appear to have noticed. Though there is legislation in Brazil to promote animals' rights, these laws are not always followed or known by producers. Federal Decree No. 24645, article 3, in Brazil forbids producers to keep animals in anti-hygienic places, places where they cannot breath, move, or rest, or that have no light or air. Producers are also forbidden to abandon sick, wounded, mutilated or extenuated animals. Producers are required to provide veterinary assistance so as to enable animals to aid animals and to make sure that the deaths are suffering-free death whether it be for consumption or not. There are also rules governing transport of animals in terms of crowding and containment.

In Thailand, larger-scale export-oriented farms appear to be trying to comply with EU standards on animal density, number of animals per worker, number of dark hours per days, and transportation conditions. Some farms have found that following these standard to be beneficial for them as both the mortality rate and Feed Conversion Ratio (FCR) improved, though there has been no rigorous study to confirm if the average cost per bird (including fixed cost) increase or decrease (Poapongsakorn *et al.,* 2003). However, the large-scale Thai farms that mainly serve the domestic market-most of which follow the western blueprint based on capital-intensive and land-scarce production systems-might be viewed as less considerate to the animals. In layer farms, for example, animal density in an evaporative cooling house is generally higher than traditional house because of the relatively high investment cost. The higher density could make the living environment less desirable, although it is also arguably the case that adopting the cooling system would make the animals more comfortable.

However, unlike some layer farms in the US where each hen has average space that equivalent to a sheet of letter paper, a large layer farm in the Thai country study let hens wander freely within the hen-houses Many small commercial farms in our survey (of sizes 50-60 pig per farms) in Nakorn Ratchasima province also use housing practices that resemble backyard farms more than industrialised farms. In these farms, pigs are allowed to go in and out their pen at will. While many Thai exporters view the animal welfare measure requirement as protectionism, according to the Thai country study, some are rather optimistic that Thailand is in a better position to follow these guidelines than a major competitor like the US (Poapongsakorn *et al.*, 2003).

Animal Diseases

Animal diseases that people can catch are called zoonoses. About 75 percent of the new diseases affecting humans in the past decade can be traced to animals or animal products. You can get a disease directly from an animal, or indirectly, through the environment.

Farm animals can carry diseases. If you touch them or things they have touched, like fencing or buckets, wash your hands thoroughly. Adults should make sure children who visit farms or petting zoos wash up as well. Pets can also make you sick. Reptiles pose a particular risk. Turtles, snakes and iguanas can transmit *Salmonella* bacteria to their owners. You can get rabies from an infected dog or toxoplasmosis from handling kitty litter of an infected cat. The chance that your dog or cat will make you sick is small. You can reduce the risk by practicing good personal hygiene, keeping pet areas clean and keeping your pets' shots up-to-date.

Animal Health

The Foreign Animal Disease Diagnostic Laboratory (FADDL) is part of the only facility in the United States where many infectious foreign animal disease (FAD) agents are studied. It is located on Plum Island, which is 1.5 miles off the northeastern end of Long Island, New York. Scientists at the FADDL are devoted to diagnosing foreign diseases of animals. They partner with scientists of the Department of Homeland Security and the USDA's Agricultural Research Service, also located on Plum Island, in foreign animal disease research.

Additionally, the FADDL is the custodian of the North American Foot-and-Mouth Disease (FMD) Antigen Bank. The Bank stores concentrated FMD antigen that can be formulated into vaccines if an FMD introduction occurs. The bank is co-owned by Canada, Mexico, and the United States. Personnel working in the vaccine bank are responsible for performing safety and potency testing of new antigen lots of FMD vaccine, and periodically testing the quality of stored antigen.

The FADDL is composed of three sections: Diagnostic Services, Reagents and Vaccine Services, and the Proficiency and Validation Section. Scientists in the Diagnostic Services Section have the capability to diagnose more than 30 exotic animal diseases (including FMD, classical swine fever, African swine fever, and other diseases listed by the World Organisation for Animal Health (OIE)), and they perform thousands of diagnostic tests each year, looking for the presence of FAD agents. Tissue and blood samples to be tested are submitted by veterinarians suspecting an exotic disease in domestic livestock or by animal import centres testing quarantined animals for foreign diseases. Samples also are submitted by animal health professionals in other countries who need help with a diagnosis.

The Reagents and Vaccine Services Section provides diagnostic reagents, assays, vaccines and other services for identification, control, and eradication of foreign animal diseases. The section is responsible for production and quality testing of reagents used at FADDL in diagnostic assays. It also participates in safety and potency evaluations of potential vaccines against foreign animal diseases.

The Proficiency and Validation Section develops and validates diagnostic technologies that can bedeployed to the Ntional Animal Health Laboratory Network (NAHLN) and/or used at the NVSL reference laboratories. The section is responsible for a training and proficiency testing program for the NAHLN and collaborates with the Department of Homeland Security National Bioforensics Analysis Centre to maintain and execute agricultural forensics capabilities. The FADDL provides training to State, Federal, and foreign officials in the clinical and laboratory diagnosis of foreign animal diseases.

The Regional Animal Disease Diagnostic Laboratory

The Regional Animal Disease Diagnostic Laboratory is located at the National Highway at Cabañgan, Camalig, Albay with a total land area of more or less 2,000 sq. metres. It is mandated to:

- Monitor the incidence and prevalence of infections/notifiable diseases present as well as newly introduced diseases and parasites in the different provinces regionwide.
- Control and eradication of animal diseases and parasites through the use of laboratory diagnostic procedures.
- Serve as reference centre for disease diagnostic methods.
- Offer livestock and poultry farmers effective services for diagnostic of bacterial, viral, and parasitic diseases as well as nutritional deficiencies and other related conditions.
- Provide technical assistance on any aspects of animal health necessary for the implementation of control and prophylactic measures for animal protection.
- Carry out researches on major diseases of domestic animals with a view of conferring practical solution for immediate

application. Assist in making decisions on policies of diseases control especially if these are to be based on knowledge of Epidemiology and economic aspects of disease control.

Animal Disease

Animal Disease, an impairment of the normal state of an animal that interrupts or modifies its vital functions. Concern with diseases that afflict animals dates from the earliest human contacts with animals and is reflected in early views of religion and magic. Diseases of animals remain a concern principally because of the economic losses they cause and the possible transmission of the causative agents to humans. The branch of medicine calld veterinary medicine deals with the study, prevention, and treatment of diseases not only in domesticated animals but also in wild animals and in animals used in scientific research. The prevention, control, and eradication of diseases of economically important animals are agricultural concerns. Programs for the control of diseases communicable from animals to man, called zoonoses, especially those in pets and in wildlife, are closely related to human health. Further, the diseases of animals are of increasing importance, for a primary public-health problem throughout the world is animal-protein deficiency in the diet of humans. Indeed, both the United Nations Food and Agricultural Organisation (FAO) and the World Helth Organisation (WHO) have been attempting to solve the problem of protein deficits in a world whose human population is rapidly expanding.

General Considerations

Historical evidence, like that frm currently developing nations, indicates that veterinary medicine originally developed in response to the needs of pastoral and agricultural man along with human medicine. It seems likely that a veterinary profession existed throughout a large area of Africa and Asia from at least 2000 bc. Ancient Egyptian literature includes monographs on both animal and human diseases. Evidence of the parallel development of human and veterinary medicine is found in the writings of Hippocrates on medicine and of Aristotle, who described the symptomatology and therapy of the diseases of animals, includin man.

Early Greek scholars, noting the similarities of medical problems among the many animal species, taught both human and veterinary medicine. In the late 4th century bc, Alexander the Great designed programs involving the study of animals, and medical writings of the Romans show that some of the most important early observations on the natural history of disease were made by men who wrote chiefly about agriculture, particularly the aspect involving dmesticated animals. Most of the earliest suggestions of relationships between human health and animal diseases were part of folklore, magic, or religious practice. The Hindu's concern for the well-being of animals, for example, originated in his belief in reincarnation. Era to about 1500, the distinctions between the practices of human and veterinary medicine were not clear-cut; this was especially true in the fields of obstetrics and orthopedics, in which animal doctors in rural areas often delivered babies and set human-bone fractures. It was realised, however, that training in one field was inadequate for practicing in the other, and the two fields were separated.

Veterinary literature from the civilizations of Greece and Rome contains reference to "herd factors" in disease; contagion within groups of animals kept together, therefore, was recognised, and both quarantine and slaughter were used to control outbreaks of livestock diseases. Rinderpest (cattle plague) was the most important livestock disease from the 5th century until control methods were developed. Serious outbreaks of the disease prompted the founding of the first veterinary college (École Nationale Vétérinaire), in Lyon, France, in 1762. Many aspects of animal diseases are best understood in terms of population or herd phenomena; for example, herds of livestock, rather than individual animals, are vaccinated against specific diseases, and housing, nutrition, and breeding practices are related to the likelihood of illness in the herd.

The work of Pasteur was of fundamental significance to general medicine and to agriculture. Veterinarians became concerned with foods of animal origin after the discovery of microorganisms and their identification with diseases in man and other animals. Efforts were directed toward protecting humans from diseases of animal origin, primarily those transmitted through meat or dairy products. Modern principles of food hygiene, first established for the dairy and meat-packing industries in the 19th and early 20th centuries, have been generally applied to other food-related industries. The veterinary profession, especially in Europe, assumed a major role in early food-hygiene programs. Since World War II, the eradication of animal diseases, rather than their control, has become increasingly important, and conducting basic research, combatting zoonoses, and contributing to man's food supply have become indispensable services of veterinary medicine.

Economic Importance

About 50 percent of the world's population suffers from chronic malnutrition and hunger. Inadequate diet claims many thousands of lives each day. When the lack of adequate food to meet present needs for an estimated world population of more than 4,600,000,000 in the 1980s is coupled with the prediction that the population may increase to 7,000,000,000 by the year 2000, it becomes obvious that animal-food supplies must be increased. Control the diseases that afflict animals throughout the world, especially in the developing nations of Asia and Africa, where the population is expanding most rapidly. Most of the information concerning animal diseases, however, applies to domesticated animals such as pigs, cattle, and sheep, which are relatively unimportant as food sources in these nations. Remarkably little is known of the diseaes of the goat, the water buffalo, the camel, the elephant, the yak, the llama, or the alpaca; all are domesticated animals upon which the economies of many developing countries depend. It is in these countries that increased animal production resulting from the development of methods for the control and eradication of diseases affecting these animals is most urgently needed.

Despite the development of various effective methods of disease control, substantial quantities of meat and milk are lost each year throughout the world. In countries in which animal-disease control is not yet adequately developed, the loss of animal protein from disease is about 30 to 40 percent of the quantity available in certain underdeveloped areas. In addition, such countries also suffer losses resulting from poor husbandry practices.Animals have long been recognised as agents of human disease. Man has probably been bitten, stung, kicked, and gored by animals for as long as he has been on earth; in addition, early man sometimes became ill or died after eating the flesh of dead animals. In more recent times, man has discovered that many invertebrate aimals are capable of transmitting causative agents of disease from man to man or from other vertebrates to man. Such animals, which act as hosts, agents, and carriers of disease, are important in causing and perpetuating human illness.

Because about three-fourths of the important known zoonoses are associated with domesticated animals, including pets, the term zoonoses was originally defined as a group of diseases that man is able to acquire from domesticated animals. But this definition has been modified to include all human diseases (whether or not they manifest themselves in all hosts as apparent diseases) that are acquired from or transmitted to any other vertebrate animal. Thus, zoonoses are naturally occurring infections and infestations shared by man and other vertebrates.

Although the role of domesticated animals in many zoonoses is understood, the role of the numerous species of wild animals with which man is less intimately associated is not well understood. The discovery that diseases suh as, viral brain infections, plague, and numerous other important diseases involving man or his domesticated animals are fundamentally diseases of wildlife and exist independently of man and his civilization, however, has increased the significance of studying the nature of wildlife diseases. Table 10 contains a partial list of zoonoses, including the causative agents and the animals involved.

Methods of Examination

Before an unhealthy animal receives treatment, an attempt is made to diagnose the disease. Both clinical findings, which include symptoms that are obvious to a nonspecialist and clinical signs that can be appreciated only by a veterinarian, and laboratory test results may be necessary to establish the cause of a disease. A clinical examination should indicate if the animal is in good physical condition, is eating adequately, is bright and alert, and is functionin in an apparently normal manner. Many disease processes are either inflammatory or result from tumours. Malignant tumours (*e.g.,* melanomas in horses, squamous cell carcinomas in small animals) tend to spread rapidly and usually cause death. Other diseases cause the circulatory disturbances or the degenerative and infiltrative changes that are summarised in the preceding section. If a specific diagnosis is not possible, the symptoms of the animal are treated.

It includes a description of the animal (age, species, sex, breed); the owner's report; the animal's history; a description of the preliminary examination; clinical findings resulting from an examination of body systems; results of specific laboratory tests; diagnosis regarding a specific cause for the disease (etiology); outlook (prognosis); treatment; case progress; termination; autopsy, if performed; and the utilisation of scientific references, if applicable.

The veterinarian must diagnose a disease on the basis of a variety of examinations and tests, since he obviously cannot interrogate the animal. Methods used in the preparation of a diagnosis include inspection—a visual examination of the animal; palpation—the application of firm pressure with the fingers to tissues to determine characteristics such as abnormal shapes and possible tumours, the presence of pain, and tissue consistency; percussion—the application of a short, sharp blow to a tissue to provoke an audible response from body parts directly beneath; auscultation—the act of listening to sounds that are produced by the body during the performance of functions (*e.g.,* breathing, intestinal movements); smells—the recognition of characteristic odours associated with certain diseases; and miscellaneous diagnostic procedures, such as eye examinations, the collection of urine, and heart, esophageal, and stomach studies.

General Inspection

Deviation of various characteristics from the normal, observation of which constitutes the general inspection of an animal, is a useful aid in diagnosing disease. The general inspection includes examination of appearance; behaviour; body condition; respiratory movements; state of skin, coat, and abdomen; and various common actions.

The appearance of an animal may be of diagnostic significance; small size in a pig may result from retardation of growth, which is caused by hog-cholera virus. Observation of the behaviour of an animal is of value in diagnosing neurological diseases; *e.g.,* muscle spasms occur in lockjaw (tetanus) in dogs, nervousness and convulsions in dogs with distemper, dullness in horses with equine viral encephalitis, and excitement in animals suffering from lead poisoning. Subtle behavioural changes may not be noticeable. The general condition of the body is of value in diagnosing diseases that cause excessive leanness (emaciation), including certain cancers, or other chronic diseases, such as a deficiency in the output of the adrenal glands or tuberculosis. Defective teeth also may point to malnutrition and result

The respiratory movements of an animal are important diagnostic criteria; breathing is rapid in young animals, in small animals, and in animals whose body temperature is higher than normal. Specific respiratory movements are characteristic of certain diseases—*e.g.,* certain movements in horses with heaves (emphysema) or the abdominal breathing of animals suffering from painful lung diseases.

The appearance of the skin and hair may indicate dehydration by lack of pliability and lustre; or the presence of parasites such as lice, mites, or fleas; or the presence of ringworm infections and allergic reactions by the skin changes they cause. The poisoning of sheep by molybdenum in their hay may be diagnosed by the loss of colour in the wool of black sheep. Distension of the abdomen may indicate bloat in cattle or colic in horses. Abnormal activities may have special diagnostic meaning to the veterinarian. Straining during urination is associated with bladder stones; increased frequency of urination is associated with kidney disease (nephritis), bladder infections, and a disease of the pituitary gland (diabetes insipidus).

Excessive salivation and grinding of teeth may be caused by an abnormality in the mouth. Coughing is associated with pneumonia. Some diseases cause postural changes: for example, a horse with tetanus may stand in a stiff manner. An abnormal gait in an animal made to move may furnish evidence as to the cause of a disease, as louping ill in sheep.

Clinical Examination

Following the general inspection of an animal thought to have contracted a disease, a more thorough clinical examination is necessary, during which various features of the animal are studied. These include the visible mucous membranes (conjunctiva of the eye, nasal mucosa, inside surface of the mouth, and tongue); the eye itself; and such body surfaces as the ears, horns (if present), and limbs. In addition, the pulse rate and the temperature are measured.

The veterinarian examines the visible mucous membranes of the eye, nose, and mouth to determine if jaundice, hemorrhages, or anemia are present. The conjunctiva, or lining of the eye, may exhibit pus in pinkeye infections, have a yellow appearance in jaundice, or exhibit small hemorrhages in certain systemic diseases. Examination of the nose may reveal ulcers and vesicles (small sacs containing liquid), as in foot-and-mouth disease, a viral disease of cattle, or vesicular exanthema, a viral disease of swine. Ulceration of the tongue may be apparent in animals suffering from actinobacillosis

A detailed examination of the eye may show abnormalities of the cornea resulting from such diseases as infectious hepatitis in dogs (Table 5), bovine catarrhal fever, and equine influenza. Cataract, a condition in which the passage of light through the lens of the eye is obstructed, may result from a disorder of carbohydrate metabolism (diabetes mellitus), infections, or a hereditary defect.

An elevated temperature, or fever, resulting from the multiplication of disease-causing organisms may be the earliest sign of disease. The increase in temperature activates the body mechanisms that are necessary to fight off foreign substances. Measuring the pulse rate is useful in determining the character of the heartbeat and of the circulatory system.

Tests as Diagnostic Aids

In many cases, the final diagnosis of an animal disease is dependent upon a laboratory test. Some involve measuring the amount of certain chemical constituents of the blood or body fluids, determining the presence of toxins (poisons), or examining the urine and feces. Other tests are designed to identify the causative agents of the disease. The removal and examination of tissue or other material from the body (biopsy) is used to diagnose the nature of abnormalities such as tumours. Specific skin tests are used to confirm the diagnoses of various diseases—*e.g.,* tuberculosis and Johne's disease in cattle and glanders in horses.

Confirmation of the presence in the blood of abnormal quantities of certain constituents aids in diagnosing certain diseases. Abnormal levels of protein in the blood are associated with some cancers of the bone, such as multiple myeloma in horses and dogs. Animals with diabetes mellitus have a high level of the carbohydrate glucose and the steroid cholesterol in the blood. The combination of an increase in the blood level of cholesterol and a decrease in the level of iodine bound to protein indicates hypothyroidism (underactive thyroid gland). A low level of calcium in the serum component of blood confirms milk fever in lactating dairy cattle. An increase in the activities of certain enzymes (biological catalysts) in the blood indicates liver damage. An increase in the blood level of the bile constituent bilirubin is used as a diagnostic test for hemolytic crisis, a disease in which red blood cells are rapidly destroyed by organisms such as *Babesia* species in dogs and in cattle and *Anaplasma* species in cattle.

The examination of the formed elements of blood, including the oxygen-carrying red blood cells (neutrophils, eosinophils, basophils, lymphocytes, and monocytes), and the platelets, which function in blood coagulation, is helpful in diagnosing certain diseases. Examination of the blood cells of cattle may reveal abnormal lymphocytic cells characteristic of leukemia. Low numbers of leucocytes indicate the presence of viral diseases, such as hog cholera and infectious hepatitis in dogs. Neutrophil levels increase in chronic bacterial diseases, such as canine pneumonia and uterine infections in female animals. Elevated monocyte levels occur in chronic granulomatous diseases; *e.g.,* histoplasmosis and tuberculosis. Canine parasitism and allergic skin disorders are characterised by elevated eosinophil levels. Prolonged clotting time may be associated with a deficiency of platelets.

Anemia has many causes. They include hemorrhages from blood loss after injuries; the destruction of red blood cells by the rickettsia *Haemobartonella felis* in cats; incompatible blood transfusions in dogs; the inadequate production of normal red blood cells, which occurs in iron or cobalt deficiency after exposure to radioactive substances; general malnutrition; and contact with substances that depress the activity of bone marrow.

Poisonings occur commonly in animals. Some species are more sensitive to certain poisons than others. Swine develop mercury poisoning if they eat too much grain that has been treated with mercury compounds to retard spoilage. Dogs may be poisoned by the arsenic found in pesticides or by strychnine, which is found in rat poison. Many plants are poisonous if eaten, such as bracken fern, which poisons cattle and horses, and ragwort, which contains a substance poisonous to the liver of cattle.

Examination of an animal's urine may reveal evidence of kidney diseases or diseases of the entire urinary system or a generalised systemic disease. The presence of protein in the urine of dogs indicates acute kidney disease (nephritis). Although constituents of bile normally are found in the urine of dogs, the quantity increases in dogs with the presence of infectious hepatitis, a disease of the liver. The presence of abnormal amounts of the simple carbohydrate glucose and of ketone bodies (organic compounds involved in metabolism) in an animal's urine is used to diagnose diabetes mellitus, a disease in which the pancreas cannot form adequate quantities of a substance (insulin) important in regulating carbohydrate metabolism. The urine of horses with azoturia (excessive quantities of nitrogen-containing compounds in the urine) or muscle breakdown may contain a dark-coloured molecule called myoglobin.

The presence of eggs or parts of worms in the excrement of animals suspected of suffering from intestinal parasites, such as roundworms, tapeworms, or flatworms, aids in diagnosis. Feces that are light in colour, have a rancid odour, contain fat, and are poorly formed may indicate the existence of a chronic disease of the pancreas. Clay-coloured fatty feces suggest obstruction of the bile duct, which conveys bile to the intestine during digestion.

The identification of a disease-causing microorganism within an animal enables the veterinarian to choose the best drug for therapy. Agglutination tests, which utilise serum samples of animals and microorganisms suspected of causing a disease, many times confirm the presence of the following bacterial diseases: brucellosis in cattle and swine, salmonellosis in swine, leptospirosis in cattle, and actinobacillosis in swine and cattle. Other tests measure the antibodies (specific proteins formed in response to a foreign substance in the body) formed against a disease-causing agent, such as those that cause brucellosis, foot-and-mouth disease, infectious hepatitis in dogs, and fowl pest.

The modern veterinary diagnostic laboratory performs, in addition to the tests mentioned, tests of cells in the bone marrow; specific-organ-function tests (liver, kidney, pancreas, thyroid, adrenal, and pituitary glands); radioisotope tests, tissue biopsies, and histochemical analyses; and tests concerning blood coagulation and body fluids.

Survey of Animal Diseases

Infectious and Noninfectious Diseases

Diseases may be either infectious or noninfectious. The term infection, as observed earlier, implies an interaction between two living organisms, called the host and the parasite. Infection is a type of parasitism, which may be defined as the state of existence of one organism (the parasite) at the expense of another (the host). Agents (*e.g.*, certain viruses, bacteria, fungi, protozoans, worms, and arthropods) capable of producing disease are pathogens. The term pathogenicity refers to the ability of a parasite to enter a host and produce disease; the degree of pathogenicity—that is, the ability of an organism to cause infection—is known as virulence. The capacity of a virulent organism to cause infection is influenced both by the characteristics of the organism and by the ability of the host to repel the invasion and to prevent injury. A pathogen may be virulent for one host but not for another. Pneumococcal bacteria, for example, have a low virulence for mice and are not found in them in nature; if introduced experimentally into a mouse, however, the bacteria overwhelm its body defences and cause death.

Pathogens enter the body in various ways—by penetrating the skin or an eye, by being eaten with food, or by being breathed into the lungs. After their entry into a host, pathogens actively multiply and produce disease by interfering with the functions of specific organs or tissues of the host. Table 3 lists some infectious and parasitic diseases of animals and the causative agents.

Before a disease becomes established in a host, the barrier known as immunity must be overcome. Defence against infection is provided by a number of chemical and mechanical barriers, such as the skin, mucous membranes and secretions, and components of the blood and other body fluids. Antibodies, which are proteins formed in response to a specific substance (called an antigen) recognised by the body as foreign, are another important factor in preventing infection. Immunity among animals varies with species, general health, heredity, environment, and previous contact with a specific pathogen.

As certain bacterial species multiply, they may produce and liberate poisons, called exotoxins, into the tissues; other bacterial pathogens contain toxins, called endotoxins, which produce disease only when liberated at the time of death of the bacterial cell. Some bacteria, such as certain species of *Clostridium* and *Bacillus*, have inactive forms called spores, which may remain viable (*i.e.,* capable of developing into active organisms) for many years; spores are highly resistant to environmental conditions such as heat, cold, and chemical compounds called disinfectants, which are able to kill many active bacteria.

The term infestation indicates that animals, including spiny-headed worms (Acanthocephala), roundworms (Nematoda), flatworms (Platyhelminthes), and arthropods such as lice, fleas, mites, and ticks, are present in or on the body of a host. An infestation is not necessarily parasitic. Table 3 includes various infestations.

Noninfectious diseases are not caused by virulent pathogens and are not communicable from one animal to another. They may be caused by hereditay factors or by the environment in which an animal lives. Many metabolic diseases are caused by an unsuitable alteration, sometimes brought about by man, in an animal's genetic constitution or in its environment. Metabolic diseases usually result from a disturbance in the normal balance of the physiological mechanisms that maintain stability, or homeostasis. Examples of metabolic diseases include overproduction or underproduction of hormones, which control specific body processes; nutritional deficiencies; poisoning from such agents as insecticides, herbicides, fluorine, and poisonous plants; and inherited deficiencies in the ability to synthesize active forms of specific enzymes, which are the proteins that control the rates of chemical reactions in the body.

Excessive inbreeding (*i.e.*, the mating of related animals) among all domesticated animal species has resulted in an increase in the number of metabolic diseases and an increase in the susceptibility of certain animals to infectious diseases.

As stated previously, zoonoses are human diseases acquired from or transmitted to any other vertebrate animal. Zoonotic diseases are common in currently developing countries throughout the world and constitute, with starvation, the major threat to human health. More than 150 such diseases are known; some examples are listed in Table.

Zoonoses may be separated into four principal types, depending on the mechanisms of transmission and epidemiology. One type includes the direct zoonoses, such as rabies and brucellosis, which are maintained in nature by one vertebrate species. The transmission cycle of the cyclozoonoses, of which tapeworm infections are an example, requires at least two different vertebrate species. Both vertebrate and invertebrate animals are required as intermediate hosts in the transmission to humans of metazoonoses; arboviral and trypanosomal diseases are good examples of metazoonoses. The cycles of saprozoonoses (for example, histoplasmosis) may require, in addition to vertebrate hosts, specific environmental locations or reservoirs.

Most animals that serve as reservoirs for zoonoses are domesticated and wild animals with which man commonly associates. People in occupations such as veterinary medicine and public health, therefore, have a greater exposure to zoonoses than do those in occupations less closely concerned with animals.

In addition to the numerous human diseases spread by contact with the parasitic worm helminth and by contact with arthropods, many diseases are transmitted by the bites and venom of certain animals; poisonous or diseased food animals also transmit diseases. Dog bites may seriously injure tissues and also can transmit bacterial infections and rabies, a disease of viral origin. The bite of a diseased rat may transmit any of several diseases to man, including plague, salmonellosis, leptospirosis, and rat-bite fevers. Cat-scratch disease may be transmitted through cat bites, and the deadly herpes B virus can spread by monkey bites. The bites of venomous snakes and fish account for considerable human discomfort and death.

About 200 of the 2,500 known species of snakes can cause human disease. One estimate for snakebite deaths worldwide is 30,000 to year, the vast majority of them in Asia. Poisonous wild animals inadvertently used for food include animals harbouring the anthrax bacillus and those containing the causative agents of salmonellosis, trichinosis, and fish-tapeworm infection. The flesh of various types of fish is toxic to man. Japanese puffers, for example, contain the poisonous chemical compound tetrodotoxin; scombroid fish harbour *Proteus morganii*, which causes gastrointestinal diseases; and mullet and surmullet can cause nervous disturbances.

Approaches to the control of zoonoses differ according to the type under consideration. Because the majority of direct zoonoses and cyclozoonoses and some saprozoonoses are most effectively controlled by techniques involving the animal host, methods used to combat these diseases are almost entirely the responsibility of veterinary medicine. A good example is the elimination of stray dogs, for they are an important factor in the control of zoonoses such as rabies, hydatid disease, and visceral larva migrans. In addition, the control of diseases such as brucellosis and tuberculosis in cattle involves a combination of methods—mass immunisation, diagnosis, slaughter of infected animals, environmental disinfection, and quarantine. Several supportive measures for the control of disease are useful in some cases. Air-sanitation measures are helpful in direct zoonoses in which human illness is spread by droplets or dust, and zoonotic infections that are spread through a fluid medium, such as water or milk, sometimes can be controlled. Heat, cold, and irradiation are effective in killing the immature forms of *Trichinella spiralis*, the causative agent of trichinosis, in meat; and certain antibiotic drugs help to prevent deterioration of food.

The control of metazoonoses may be directed at the infected vertebrate hosts, at the infected invertebrate host, or at both. Particularly effective in this instance has been the use of chemical insecticides to attack the invertebrate carriers of specific infections, even though several difficulties have been encountered—for example, the inaccessibility of the invertebrate to the chemicals, which occurs with organisms that breed in swiftly flowing waters or in dense vegetation, and the development of insecticide resistance by the organisms. Insecticides are used to destroy the mosquitoes that spread malaria (*Anopheles*). Mechanical filters placed across irrigation ditches help to prevent the dissemination of the snails that transmit *Schistosoma mansoni*, a parasitic flatworm.

Disease Prevention, Control, and Eradication

Prevention is the first line of defence against disease. At least four preventive techniques are available for use in the prevention of disease in an animal population. One is the exclusion of causative A second preventive tool utilises control methods such as immunisation, environmental control, and chemical agents to protect specific animal populations from endemic diseases, diseases normally present in an area. The third preventive measure concerns the mass education of people about disease prevention. Finally, early diagnosis of illness among members of an animal population is important so that disease manifestations do not become too severe and so that affected animals can be more easily managed and treated.

Quarantine—the restriction of movement of animals suffering from or exposed to infections such as bluetongue and scrapie (in sheep), foot-and-mouth disease (in cattle), and rabies (in dogs)—is one of the oldest tools of preventive medicine. It was applied to domesticated animals as early as Roman times. The establishment of international livestock quarantine in the United States in 1890 provided for the holding of all imported cattle, sheep, and swine at the port of entry for 90, 15, and 15 days, respectively. In this way, such diseases as Nairobi sheep disease, surra, and infections caused by *Brucella melitensis* were eliminated or excluded from the United States, but international quarantine barriers did not prevent the entry of bluetongue, scrapie, and the tick *Rhipicephalus evertsi*, which is a carrier for several animal diseases. On the other hand, long-term quarantine of all dogs entering Great Britain has been effective since its initiation in 1919 (the quarantine also includes cats). It is possible that aircraft may pose new problems regarding livestock-disease quarantine since many disease carriers (*e.g.*, insects and viruses) may be accidentally brought by plane into a country.

Mass immunisation as a preventive technique has the advantage of allowing the resistant animal freedom of movement, unlike environmental control, in which the animal is confined to the controlled area; immunisation may, however, provide only short-lived and partial protection. Mass-inoculation techniques against diseases such as Newcastle disease in chickens and distemper in mink and dogs have been successful. Animal diseases have been prevented by methods involving environmental control, including the maintenance of safe water supplies, the hygienic disposal of animal excrement, air sanitation, pest control, and the improvement of animal housing. One specific environmental program, called the portable-calf-pen system, involves routine movement of the pens to avoid a concentration of specific pathogens in them.

Other programs involve the utilisation of automatic and sanitary watering and feeding equipment and buildings with environmental controls The use of chemical compounds to prevent illness (chemoprophylaxis) includes a variety of pesticides, which are used to kill insects that transmit diseases, and substances either used internally or applied to the animal's body to prevent the transmission or the development of a disease. An example is the use of sulfonamide drugs in the drinking water of poultry to prevent coccidiosis. Environmental-control methods in the poultry industry have resulted in the most efficient means of poultry production developed thus far.

The early detection of a disease in a population of animals—a herd of cattle, for example—is particularly useful in controlling certain chronic infectious diseases, such as mastitis, brucellosis, and tuberculosis, as well as certain noninfectious diseases such as bloat. Laboratory tests—the agglutination test in pullorum disease, the tuberculin skin test for tuberculosis, the examination of feces for eggs of specific parasites, the physical and chemical tests performed on milk to diagnose bovine mastitis—are used for the early detection of diseases in an animal population.

Methods of disease control and eradication have been successful in various countries. In the United States, for example, the test-and-slaughter technique, in which simple tests are used to confirm the existence of diseased animals that are then slaughtered, has been of great value in controlling infectious and hereditary diseases, including dourine, a venereal disease in horses, fowl plague, and foot-and-mouth disease in cattle and deer. Bovine tuberculosis has been eliminated from Denmark, Finland, and The Netherlands and reduced to a low level in various other countries, including Great Britain, Japan, the United States, and Canada, by the test-and-slaughter method. Many infectious diseases have been eradicated from Great Britain—sheep pox, rinderpest, pleuropneumonia, glanders, and rabies. Dseases eliminated from Australia by a combination of methods—control of agents that carry disease, the test-and-slaughter technique, the use of chemical agents, and, more recently, biological control—include hog cholera, rinderpest, scrapie, glanders, surra, rabies, and foot-and-mouth disease.

In biological control, enemies of the agents that transmit the disease, enemies of the reservoir host, or a specific parasite are introduced into the environment. If a natural enemy of the tsetse fly could be found, for example, African sleeping sickness in man and trypanosomiasis in cattle could be controlled in West Africa. Successful biological control of the European-rabbit population in Australia has been accomplished through the use of the myxomatosis virus, which is transmitted by moquitoes and causes the formation of Although the Brazilian white rabbit is relatively unaffected by the virus, it causes rapid death in the European rabbit. The elimination of the European rabbit in France by the virus was accompanied by a decrease in tick-borne typhus in people, suggesting that the rabbit may be a significant intermediate host for the causative agent, *Rickettsia conorii*. Screwworms, an immature form of the fly *Cochliomyia hominivorax*, have been eradicated in the United States by the release of more than 3,000,000,000 sterilised males.

Disease control and elimination programs require many sophisticated techniques, in addition to diagnosis and the slaughter of affected animals. They include: the control of insects known to transmit diseases; the cooperation of animal owners; the development through research of new diagnostic tests for use on large populations; the eradication of animal species from areas in which they are known to transmit disease; sterilisation of strains of animals known to carry inheritable metabolic diseases; and effective meat inspection.

Chapter 3

Livestock Diseases—Diagnostic Service Program

Declaration of Purpose

The production of livestock is one of the largest industries in this state; and whereas livestock disease constitutes a constant threat to the public health and the production of livestock in this state; and whereas the prevention and control of such livestock diseases by the state may be best carried on by the establishment of a diagnostic service program for livestock diseases; therefore it is in the public interest and for the purpose of protecting health and general welfare that a livestock diagnostic service program be established.

Director Authorised to Carry on Diagnostic Program

The director of agriculture is hereby authorised to carry on a diagnostic service program for the purpose of diagnosing any livestock disease which affects or may affect any livestock which is or may be produced in this state or otherwise handled in any manner for public distribution or consumption.

Employment of Personnel

In carrying out such diagnostic service program the director of agriculture may employ, subject to the state civil service act, chapter 41.06 RCW, the necessary personnel to properly effectuate such diagnostic service program.

Acceptance of Gifts, Funds, Equipment, etc.

In carrying out such diagnostic service program, the director of agriculture may accept public or private funds, gifts or equipment or any other necessary properties.

Schedule of Fees may be Established—Use

The director may, following a public hearing, establish a schedule of fees for services performed in carrying out such diagnostic service program. All fees collected under this provision shall be retained by the director of agriculture to be spent only for carrying out the purposes of this chapter.

Disposal of Dead Animals

Definitions

For the purposes of this chapter, unless clearly indicated otherwise by the context:

(1) "Director" means the director of agriculture;

(2) "Meat food animal" means cattle, horses, mules, asses, swine, sheep and goats;

(3) "Dead animal" means the body of a meat food animal, or any part or portion thereof: Provided, That the following dead animals are exempt from the provisions of this chapter:

- (a) Edible products from a licensed slaughtering establishment;
- (b) Edible products where the meat food animal was slaughtered under farm slaughter permit;
- (c) Edible products where the meat food animal was slaughtered by a bona fide farmer on his own ranch for his own consumption;
- (d) Hides from meat food animals that are properly identified as to ownership and brands;

(4) "Carcass" means all parts, including viscera, of a dead meat food animal;

(5) "Person" means any individual, firm, corporation, partnership, or association;

(6) "Rendering plant" means any place of business or location where dead animals or any part or portion thereof, or packing house refuse, are processed for the purpose of obtaining the hide, skin, grease residue, or any other byproduct whatsoever;

(7) "Substation" means a properly equipped and authorised concentration site for the temporary storage of dead animals or packing house refuse pending final delivery to a licensed rendering plant;

(8) "Place of transfer" means an authorised reloading site for the direct transfer of dead animals or packing house refuse from

the vehicle making original pickup to the line vehicle that will transport the dead animals or packing house refuse to a specified licensed rendering plant;

(9) "Independent collector" means any person who does not own a licensed rendering plant within the state of Washington but is properly equipped and licensed to transport dead animals or packing house refuse to a specified rendering plant.

Duty to Bury Carcass of Diseased Animal—Dead Animal Presumed Diseased

Every person owning or having in charge any animal that has died or been killed on account of disease shall immediatly bury the carcass thereof to such a depth that no part of the carcass shall be nearer than three feet from the surface of the ground. Any animal found dead shall be presumed to have died from and on account of disease.

Sale, Gift, or Conveyance Prohibited—Exceptions

It is unlawful for any person to sell, offer for sale or give away a dead animal or convey the same along any public road or land not his own: Provided, That dead animals may be sold or given away to and legally transported on highways by a person having an unrevoked, annual license to operate a rendering plant or by a person having an unrevoked, annual license to operate as an independent collector.

License Required of Rendering Plants and Independent Collectors

It is unlawful for any person to operate a rendering plant or act as an independent collector without first obtaining a license from the director.

Right of Access to Premises and Records

The director or his authorised agent, shall have free and uninterrupted access to all parts of premises that come under the provisions of this chapter, for the purpose of making inspections and the examination of records.

Unlawful Possession of Horse Meat—Exceptions

It shall be unlawful for any person to transport, to sell, offer to sell, or have on his premises horse meat for other than human consumption unless said horse meat is decharacterised in a manner prescribed by the director: Provided, That this provision shall not apply to carcasses slaughtered by a farmer for consumption on his own ranch or to carcasses in the possession of a person licensed under this chapter, or to canned horse meat meeting United States bureau of animal industry regulations.

Medical Milestones

Thanks to research utilising animals, many diseases that once killed or disabled millions of people every year — such as polio, diphtheria, mumps, rubella and hepatitis — are preventable or treatable, or perhaps have been eradicated altogether. The use of animals in research studies has helped scientists to develop procedures as blood transfusion, dialysis and organ-transplantation. In fact, since 1979, nearly every Nobel Prize awarded in Medicine has been for research which depended on laboratory animal models.

Here at UVM, the use of animal models has been crucial to many areas of research. A few recent studies that implement animal studies are:

Developing Targeted Cancer Treatments

Researchers at UVM are hard at work developing "targeted" chemotherapeutic agents for cancer. Targeted chemotherapeutic agents may be more effective than current treatments at fighting cancer, without the side effects.

How are they doing this? Cancer researchers use patients' tumour cells (collected by biopsy). These tumour cells are used to establish the same tumour in immunodeficient mice. Because the mice lack key components of their immune system, they do not recognise the cells as being foreign, and hence the tumour will grow in the mice. The mice can then be used to ascertain the effectiveness of different chemotherapeutic drugs, or combinations of drugs, or to look at novel therapies such as antibodies "personalised" against a patient's specific tumour cell-surface antigens.

Providing New Hope for People with Anxiety Disorders

Anxiety usedto be understood only as a product of fear, within the limited context of negative emotions. Now, a cross-disciplinary team of UVM researchers — spanning the departments of Neurobiology, Psychology and Chemistry — is investigating a key region of the brain and their work holds great promise for people affected by anxiety disorders.

How are they doing this? The research team is using rats to understand the impact of a sometimes-neurotransmitter called pituitary adenylate cyclase-activating peptide, or PACAP, in a little-studied region of the brain — the bed nucleus of the stria terminalis. Already the scientists have been able to block anxiety behaviours in rats through the use of a series of small organic molecules that block

PACAP. Their goal: to find the first truly targeted medications for chronic anxiety diseases from anorexia to post-traumatic stress disorder.

Perfecting Non-Invasive Techniques for Treating Heart Disease

The number one cause of death in the United States is heart disease. UVM researchers are developing treatments for atheroscl erois and looking for ways to heal the heart muscle.

How are they doing this? Some genetically modified strains of mice carry genes that are defective in the biochemical pathways that process cholesterl. These mice develop atherosclerosis in their coronary arteries, similar to humans. UVM researchers have used these animals to screen novel drugs for the reduction of atherosclerotic lesions prior to utilising the drugs in human trials. Cardiology investigators also have developed a method for surgically inducing a myocardial infarction ("heart attack") in mice. These mice then can be used to test therapies which will improve the healing capacity of the heart muscle, from new drugs to stem cell-mediated treatments.

Looking at Learning

Pioneering work in UVM's Department of Psychology — connecting neural events and human behaviour — has made our researchers in-demand speakers and contributors to journals where they translate basic experiments into insights that maylead to better treatment for issues like anxiety disorders and drug addiction.

How are they doing this? Extinction learning — the process, for example, of teaching a dog that a bell no longer means food — inhibits original learning but doesn't erase it. Our researchers have discovered that extinction learning is dependent on the context or environment in which it is learned. So when a person leaves the therapist's office or rehab clinic, the lurking habit, fear, or addiction becomes the default, explaining the prevalence of relapse. Using rats, researchers space exposure trials at different intervals, teaching the rats a rhythm of sorts, then test when or whether the rats achieve extinction.

Creating New Treatments to Control Diabetes

Type 1 diabetes affects one in every 600 children and can shorten a person's lifespan by up to 15 years. It's not a disease sufferers can outgrow and controlling it requires constant vigilance. That's why endocrinology researchers at UVM are working on a treatment that ends the need for injections.

How are they doing this? Researchers utilise mice to exaine the factors that promote the growth of beta-islet cells — the cells that secrete insulin. When a portion of the pancreas is surgically removed in mice, the islet cells in the remaining pancreas compensate by increasing in number and metabolic activity. The use of genetically modified mice, which lack specific protein intermediaries, has helped to demonstrate how the growth of these cells is regulated. In the future, this knowledge will allow the treatment of type 1 diabetes by stimulating the patient's own beta-islet cell development or performing islet-cell transplants, rather than relying on insulin injections.

Providing Better Health for People with Cystic Fibrosis

Patients with cystic fibrosis — an inherited chronic disease — are often artificially ventilated for long periods of time. Artificial ventilation results in an increased risk of lung infection that is resistant to antibiotics. UVM researchers are investigating new ways to combat this bacterium.

How are they doing this? Because the bacterium *Pseudomonas bacteria* forms biofilms, infections are particularly difficult to treat with conventional antibiotic therapy. Using a mouse model of Pseudomonas pneumonia, UVM researchers in Microbiology and Molecular Genetics are investigating novel non-antibiotic substances which appear to increase the bacteria's susceptibility to antibiotics.

Three Main Diseases Genetically Linked

The origin of three costly cattle diseases is genetically linked, according to findings from USDA researchers.

Pinkeye, foot rot and bovine respiratory disease (pneumonia) are all correlated with bovine chromosome 20, say scientists at the Agricultural Research Service's Roman L. Hruska U.S. Meat Animal Research Centre in Clay Centre, Nebraska.

Bovine respiratory disease accounts for 75% of feedlot illnesses and up to 70% of all deaths, with economic losses to cattle producers exceeding $1 billion annually.

The estimated costs for pinkeye are $150 million yearly, and losses to dairy producers due to foot rot range from $120 to $350 per animal.

Eduardo Casas, research leader of the Ruminant Diseases and Immunology Research Unit at the ARS National Animal Disease Centre in Ames, Iowa, and a former USMARC geneticist, examined the genetic makeup of cattle for evidence of genes associated with resistance or tolerance to diseases.

Casas and his colleagues combined pinkeye, foot rot and bovine respiratory disease to represent overall pathogenic disease incidence.

They developed half-sibling families from crossbreed bulls: a Brahman-Hereford, a Brahman-Angus, a Piedmontese-Angus and a Belgian Blue-MARC III, which is part Red Poll, Pinzgauer, Hereford and Angus. An analysis of DNA samples from the 240 offspring infected with one or more of the diseases revealed a genetic marker, called a quantitative trait locus, on chromoome 20. This correlation shows the quantitative trait locus is associated with the three diseases. Chromosome 20 is located near genetic markers related to other diseases and may have a significant effect on the overall health of cattle, says Casas. Identifying genetic markers responsible for disease would provide an opportunity to produce cattle with increased disease tolerance, which also could help reduce economic loss associated with diseases.

Foreign Animal Diseases Lurk as Threat

The United States is highly vulnerable to the accidental or intentional introduction of foreign or emerging animal diseases, says James Roth professor of veterinary medicine at Iowa State University.

In testimony presented July 20 to the U.S. Senate Committee on Agriculture, Nutrition and Forestry in Washington D.C., Roth said more needs to be done and he recommended that vaccines be developed. The committee held a hearing to review biosecurity preparedness and efforts to address agroterrorism. Roth is a distinguished professor in veterinary medicine and director of he Centre for Food Security and Public Health at ISU. Although progress has been made, Roth says, the U.S. continues to have an inadequate infrastructure to prevent, detect, respond or recover from an outbreak of a foreign animal or zoonotic disease. A zoonotic disease infects both animals and people.

Concerned about Threat to Humans

"Agents against animals have been considered a component of nearly every nation-sponsored offensive biowarfare program," says Roth. "In recent years, there have been numerous examples in other countries of accidental introductions of foreign animal diseases and zoonotic diseases with devastating consequences."

Many foreign animal diseases also can infect people, Roth warns, including three leading threats — Avian influenza, Rift Valley fever and Nipah virus.

Roth recommended the Senate target three urgent priorities to strengthen the country's ability to protect public health, animal health and agriculture from disease threats. These priorities are rapid development of vaccines and antivirals for high priority diseases, enhancing physical facilities for animal health research and disease diagnosis and increasing human resources in veterinary medicine and public.

Priorities Needed to Protect the Public

Roth specifically urged the Senate to shift a portion of Project Bioshield's $5.6 billion budget for human vaccines to the development of vaccines for zoonotic diseases in animals. He also recommended that steps be taken immediately to replace inadequate facilities at Plum Island Animal Disease Centre in New York.

Roth endorsed the Veterinary Medical Workforce Expansion Act, which allocates a major investment for veterinary medicine colleges to address the shortage in food animal and public health veterinarians. He also called for funding of the National Veterinary Medical Services Act, which will provide loan repayment for veterinary students who agree to work in underserved rural areas. Iowa State's Centre for Food Security and Public Health strives to increase national preparedness for accidental or intentional introduction of disease agents that threaten food security or public health.

Animal Disease Diagnosis and Research Accelerates

Agriculture Secretary Ann M. Veneman announced that President Bush would include $178 million in the FY 2005 budget to complete the renovation of the U.S. Department of Agriculture's new National Centres for Animal Health. The secretary made the announcement at a ceremonial ground breaking at the extensive USDA facility in Ames, Iowa, Jan. 13.

"When completed, the nearly 1 million square food centre will become the most modern and best-equipped animal disease research facility in the world," Veneman says. "The work that is done here is a crucial link to the overall effort to protect animal agriculture."

Facility Houses Joint USDA Efforts

The Ames complex is USDA's "flagship laboratory" for large animal research and diagnosis. It includes the National Animal Disease Centre, the National Veterinary Services Laboratory and the Centre for Veterinary Biologics. A 2001 report to Congress outlined options for updating and renovating the complex, now referred to as the USDA National Centres for Animal Health. Since that time, the Bush Administration has worked with Congress on a plan to renovate the facility.

"The request of $178 million by the President would represent the final installment of the $460 million needed to fully renovate these facilities," Veneman says. "If approved by Congress, these funds will permit us to fully complete this project by the end of 2007. We intend to use accelerated contract procedures and construction techniques to meet this schedule."

For Animals, it's Equivalent to Centres for Disease Control

Veneman says the facility is more important than ever before in the context of recent animal disease threats. For instance, the National Veterinary Services Laboratory conducted the initial tests to confirm the case of bovine spongiform encepholopathy, BSE, from a single cow in Washington state.

"Even though the ultimate confirmation was made in England, we had the confidence in our own experts at the National Veterinary Services Laboratory in order to make an immediate announcement and respond quickly," Veneman says.

When completed, Veneman says the National Centres for Animal Health would include almost 1 million square feet of thoroughly modern facilities that will be biosafe, energy-efficient and will provide state-of-the-art capabilities for research and diagnosis. It will house in a single location a critical mass of scientists who are at the top of their fields with programs across animal disease research, diagnostics and biologics making USDA better able to respond to foreign animal diseases and bioterriorism.

Within USDA, the programs and facilities of the National Centres for Animal Health are operated by the Animal and Plant Health Inspection Service and the Agricultural Research Service.

Some Statistics

The diagnostic and animal research centre will employee about 600 people. Of those, about 275 will be scientists.

The facility has about 103 buildings. Most are destined for the wrecking ball.

The new complex will have about 125 animal rooms. Each room will be able to hold 3 to 5 animals, depending on size.

Diagnostic scientists at the centre should be able to diagnose any animal disease within U.S. borders. Researchers for USDA's Agricultural Research Service will be working on 15 to 20 diseases at any given time.

Mission Differs from Plum Island

Roughly three years ago, world livestock producers focused on Foot and Mouth Disease in Europe. Extensive work done at USDA's Plum

Island, New York, Animal Disease Centre helped keep FMD out of the United States.

The Plum Island facility focuses on research to keep foreign animal diseases out of the United States. Mission of the Ames, Iowa, facility is to diagnose, contain and prevent diseases within the United States.

Tropical Medicine and Animal Diseases

The Institute demonstrated the political and economic significance attributed to the pastoral industry in South Africa and the conviction that scientific discoveries could increase output. During this period, researchers explicated the aetiology and provenance of hitherto mysterious diseases such as lamsiekte, geeldikkop and African horsesickness. They developed vaccines, some of which were adopted internationally. The nature of their investigations showed that veterinary science increasingly entailed more than just progress in biomedical procedures.

Ecological factors, in particular the nutritional state of the veld, became a priority from the 1920s onwards as veterinarians saw their function as promoting animal health as well as eliminating disease. Dealing with contagious infections also incorporated less welcome, and at times controversial, approaches to disease control. The imposition of pastoral regulations illustrated the expanding powers of the South African state, founded on presumptions of scientific legitimacy. The article also explores the contribution made by African communities and settler farmers to the institutionalisation of veterinary knowledge, as well as the role South African researchers played in the evolution of a colonial, as well as an increasingly international, scientific culture.

In January 1944, Petrus Johann du Toit, the South African Director of Veterinary Services, rejected a request from R. Parker of the United States Department of Health for strains of the heartwater *rickettsia* for which the Americans were hoping to devise and manufacture a vaccine. Du Toit refused on the grounds that 'Heartwater is an exclusively African disease and I am of opinion that the task of solving the many problems connected with this disease should in the first place devolve on African scientists'.

Du Toit described how a small group of dedicated researchers at Onderstepoort were working flat out to find a preventive and a cure for this tick-borne disease that affected sheep and calves, despite the fact that South Africa did not have the economic resources to compete on equal terms with the United States. Defending the efforts of his team, du Toit bluntly stated:

I realise of course that a scientific problem can never be any one person's preserve; that competition and even duplication of work may sometimes be very desirable; and yet I feel that in this instance I ought to protect the men who have struggled against the odds and are beginning to see daylight in a very complicated problem, which is of immense economic importance to South Africa.

Du Toit's letter is important because it says a great deal about internal perceptions of South African veterinary science in the mid-1940s. Du Toit was Director of Veterinary Services from 1927 until 1948, having succeeded the Swiss bacteriologist Arnold Theiler, who had played a key role in the founding of the Onderstepoort Veterinary Institute in 1908.

Under the direction of both Theiler and du Toit, Onderstepoort Veterinary Institute contributed considerably to the growing international body of veterinary knowledge. Significant discoveries included the aetiology of the cattle disease *lamsiekte* in the 1920s as well as the invention of an effective vaccine against African horsesickness in the 1930s. By the late 1940s, South Africa was well on the way to eradicating two major scourges of the African continent: East Coast fever, transmitted to cattle primarily by the brown tick, and nagana, conveyed to cattle and other domestic animals by the tsetse fly.

In Africa as a whole, Onderstepoort also led the way in toxicological researh into poisonous plants. During the Second World War, South African virologists provided horsesickness vaccines for the British army fighting in Egypt and the Middle East and were instrumental in promoting the anti-rinderpest inoculation campaign in Tanganyika, successfully preventing the southward spread of this cattle epizootic. Du Toit was indeed proud of South Africa's contribution to veterinary science and its relative success in controlling a number of dangerous animal diseases. In his view, South Africa could and should solve its own epidemiological problems.

Du Toit epitomised a new sense of national self-assurance that pervaded the scientific fraternity more generally. Saul Dubow has discussed this in relation to the changing attitudes amongs members of the South African Association for the Advancement of Science from the time of its founding in 1903 to the visit of their British counterparts in 1929. By the latter date South African scientists no longer saw their work as inferior or dependent upon an agenda or methodology set by the European metropole. South Africans had their own intellectual networks, peer-reviewed journals of international standing, of which the *Onderstepoort Journal* was one, as well as their own professional organisations.

In the veterinary context, the confidence of the 1940s can be contrasted with the deference to metropolitan knowledge demonstrated in 1902 when East Coast fever broke out in Southern Rhodesia and the Transvaal. This disease was new to the region and on this occasion the various southern African authorities asked the eminent German virologist, Robert Koch, to investigate. Paul Cranefield has discussed this encounter between established metropolitan scientists and those working in the colonies. Initially Koch mistakenly assumed that East Coast fever was a virulent form of another cattle disease, redwater, and like redwater was transmitted by the blue tick (*Boophilus decoloratus*). Cranefield argued that these assumptions undermined the convictions of several colonial veterinarians, who doubted Koch's conclusions, yet were hesitant about contesting his theories.

However,, in many respects East Coast fever was the turning-point for South African veterinary science, both in terms of the confidence it ultimately imparted to its scientists and in the credibility it gave them in the eyes of the public and the state. At Elands River Valley and Daspoort in the Transvaal, as well as at the Rosebank laboratories near Cape Town, Arnold Theiler and the Cape's American entomologist, Charles Lounsbury, came to different conclusions to Koch, and published their findings in 1903–04. Theiler discovered East Coast fever was not redwater but a disease caused by a different protozoan (single cell organism) that came to be known as *Theileria parva*. Lounsbury meanwhile proved that the brown tick (*Rhipichephalus appendiculatus*) was the primary vector, not the blue tick. Lounsbury's findings received international attention, being published in his home country by the American Association of Economic Entomologists, whilst the classification *Theileria parva*, granted by the metropolitan establishment, reflects the pre-eminence given to Theiler as the recognised identifier of the offending organism.

Advances in the understanding, if not the immediate eradication of East Coast fever, affirmed the importance of veterinary science as an economic pillar of the South African state and provided a justifiable basis for funding the construction of the Onderstepoort laboratories. With Union in 1910, Onderstepoort became the centre for South African veterinary research.

By 1910 South Africa had already obtained some recognition for its research in other parts of the continent. Invited as an adviser to British as well as German East Africa in 1909, Theiler described how the veterinary authorities in both colonies had adopted the South African practice of trying to tackle East Coast fever by clearing the veld of all cattle for at least a year, thereby denying the ticks their blood feed.

Although scientists later revised this advice, the idea of clean veld, which resonated with the Victorian sanitationist approach towards improving human health, constituted the latest in veterinary thinking, based as it was on entomological investigations into the life-cycle of the brown tick. Theiler's visit suggests that scientists working in other African colonies no longer automatically looked to Europe for its expertise. South Africa had become an important centre in what Roy MacLeod has described as the 'moving metropolis' of knowledge generation. Western scientific methodologies could ultimately be developed and adapted as much at the colonial periphery as at the metropole, and the colonies themselves played a significant role in the expansion of an imperial, and indeed an increasingly global, scientific culture.

The rest of this article explores several key themes that relate to the agricultural history of South Africa, as well as to debates in the history of science and medicine. It argues that veterinary science played an important role in the development of the livestock economy in South Africa. Second only to gold mining, rural products generated the largest amount of export revenue, with wool constituting the prime pastoral commodity throughout the first half of the twentieth century. The expansion of mining compounds, together with the rapid urbanisation that accompanied the manufacturing boom from the 1930s onwards, generated a growing demand for cheap meat and dairy products. The article also traces the development of veterinary research and analyses the scientific and economic reasons behind specific shifts in investigative priorities. It shows that, despite the growing understanding of the role played by germs in causing epizootics, the environment in terms of climate, the veld and its broader ecology, were key components of South Africa's disease landscape. Given that insects and ticks were responsible for transmitting a large number of diseases, scientifically South Africa fell under the generalised remit of 'tropical' medicine.

The article also highlights the importance of the local disease environment in contributing to the expansion of an international body of veterinary and ecological knowledge. Work on indigenous plants, such as *Vangueria pygmaea*, a species apparently unique to South Africa and the cause of *gousiekte* toxicosis, demonstrates a very specific focus that contrasts with studies into continent-wide diseases such as horse sickness or global problems such as botulism (*lamsiekte*). The security of South Africa's livestock industry also depended upon effective veterinary controls in neighbouring states to contain diseases such as rinderpest and nagana. The paper concludes by arguing that Onderstepoort made an important contribution to veterinary research with a relevance that extended far beyond the Limpopo.

Veterinary Science in Historical Context

Arnold Theiler believed it was the responsibility of the state to provide veterinary services and carry out research into livestock diseases that undermined the Transvaal and later the South African economy. His ideas found favour with the British government, which had taken over the former South African Republic, as well as the Orange Free State, during the South African War (1899–1902). The governor, Alfred Milner, who oversaw the early years of reconstruction, was determined to make the Transvaal a prosperous colony by promoting gold mining on the Witwatersrand, as well increasing agricultural production, primarily by white, and preferably British, settler farmers.

Milner believed in the imperial ideal that the Empire should pay for itself. Like Joseph Chamberlain, he was committed to the notion of 'constructive imperialism' by which modern science, as well as professional bureaucracies, should promote economic growth. In 1903, Milner established the Transvaal's first Department of Agriculture with the support of progressive commercial farmers and leading Afrikaner politicians such as Louis Botha and Jan Smuts. Improving the livestock economy in order to provide cheap food for the Rand as well as the cities was integral to this policy. These events mirrored the scientific and administrative approach to rural development, which had already started in the Cape with the appointment of the Colony's first state veterinarian in 1876 and the establishment of a Department of Agriculture eleven years later.

After 1910, the quest for nutritional self-sufficiency, as well as an increase in export revenues from primary products, remained a strong political ideal. Self-consciously progressive white commercial farmers were politically influential and, through organisations such as the Agricultural Union, they collectively put pressure on the state to finance research that would overcome environmental barriers to improved production.

The Department of Agriculture sponsored a variety of research programmes that looked into such issues as the development of dryland farming, the cultivation of better varieties of seeds and grasses, as well as the elimination of injurious crop pests. Nonetheless, the Veterinary Department received the largest share of state grants, which reflected not only the economic importance attributed to the livestock industry but also the high cost involved in developing this branch of science.

Many animal diseases in South Africa were unknown in Europe and, given the lack of research facilities available elsewhere in Africa at the turn of the twentieth century, South Africa-based scientists often had to investigate mysterious infections from scratch. The large distances between many farms, as well as the poverty of many farmers, militated against the emergence of a private veterinary service until the 1940s. Nor were there any pharmaceutical companies in South Africa with the capital or will to invest in inoculation and drug research. Thus, in a bid to increase pastoral yields, the state bore the cost of discovering remedies and prophylaxis, as well as enforcing regulations aimed at preventing the spread of contagious diseases.

Leading veterinarians justified this public spending by pointing to their contribution to the national economy. In 1928, for instance, Petrus du Toit argued that the €90,000 spent each year on research facilities and staffing at Onderstepoort, as well as the €300,000 needed for veterinary regulation and fieldwork, had been of immense value to the country, having saved millions of pounds worth of livestock. Between 1904 and 1929 the cattle population increased from a low of 3,500,000, in the wake of rinderpest, to 10,500,000 head, whilst the number of sheep rose from around 16,000,000 to 45,000,000 over the same period. Wool exports grew from 165,000,000 lb in 1915 to 287,000,000 lb by 1929. Du Toit also claimed that because South Africa had a harsh environment and a hot climate, diseases were much harder to tackle than in the temperate north, thereby providing an ecological dimension in his defence of veterinary spending.

At times du Toit and some of his colleagues, in particular Philip Viljoen (who was Secretary for Agriculture 1931–45), tried to boost their case by giving some support for political positions that advocated racial differentiation in the allocation of land and veterinary resources. Although there is no direct evidence to suggest that the Veterinary Department actively endorsed the 1913 Land Act and the dispossession of Africans that this entailed, du Toit did express the belief that science could open up new lands to white settlement as exemplified in the expansion of a 'beef frontier' in the Western Transvaal as well as in the land grants to ex-soldiers in tsetse-stricken Zululand after the First World War.

Du Toit was confident that veterinary science would ultimately eliminate disease, whilst the construction of boreholes and veld reclamation schemes could generate more productive pastures. Philip Viljoen meanwhile took on board the political issue of Afrikaner 'poor whiteism' and specifically linked scientific advances with the alleviation of poverty. Viljoen played an important role in drawing up the 1937 Marketing Act, intended to secure retail outlets and improve prices for settler farmers.

In defending the €90,000 spent on the Onderstepoort Veterinary Institute by the late 1920s du Toit was also reflecting upon the expanding nature of veterinary research, which was to continue throughout the first half of the twentieth century. In 1906 the Transvaal Government had bought the land at Onderstepoort, north of Pretoria, for €1,500. Theiler chose this location because of its proximity to the capital, its large acreage and its good rail networks. Environment played a role too: Onderstepoort lay in the heart of the Transvaal bushveld where horse sickness was ife and poisonous plants plentiful. The Institute's designers constructed three specialised laboratories for bacteriology, pathology and zoology, all equipped with the latest technical apparatus. Over the years they added new buildings, revealing a growing diversity in veterinary research. In this respect the increasing specialisation of individual scientists is very striking. In the early years, practitioners such as Theiler and Herbert Curson explored a variety of veterinary problems, such as vaccine manufacture as well as the bionomics of ticks and tsetse flies, veld deficienies and poisonous plants. By the 1930s, however, each researcher focused much more on a particular area of expertise. Toxicology, biochemistry, parasitology and sex physiology emerged as sub-departments, and 'germ research' was divided into virology, bacteriology and protozoology. By 1958 only 38 out of the 66 researchers at Onderstepoort had veterinary training. Although infectious diseases continued to receive the most attention, biochemistry was the largest single department, analysing the nutritional value of grasses and mineral supplements.

These developments illustrated more than just growing scientific complexities. The timing of particular specialisations was also linked to changing patterns of disease incidence as well as to new technological opportunities. At the turn of the twentieth century and before the founding of Onderstepoort, veterinarians prioritised two highly fatal bovine epizootics, rinderpest and East Coast fever. By 1905, rinderpest had been practically eliminated through vaccination and strict quarantines. East Coast fever was still a problem and illegal stock movements brought the disease into the Transkei in 1910. Gradually, however, the situation stabilised.

The disease remained confined to the Transvaal, Natal and the Transkei as regulations on stock movements and regular cattle dipping contained the infection to some degree and prevented the recurrence of widespread mass outbreaks. By 1910, vaccines were available for a number of enzootic diseases such as lung sickness and redwater in cattle, as well as bluetongue in sheep, all of which had caused considerable losses in the past. In comparative terms the second decade of the twentieth century was quiet on the disease front. There were no major epizootics to contend with, enabling researchers to investigate diseases that were not necessarily fatal but rather undermined the economic potential of the livestock industry. From around 1912 to 1930 there was a marked shift from an emphasis on germs and immunological research, executed in the laboratory, to an emphasis on the veld as the site of infection and malnutrition.

This shift to the field indicated a more environmental approach to understanding sickness as researchers sought not only pathogenic 'germs', but also analysed the wider world in which they functioned. The distribution of game and insect vectors, as well as the composition of the veld, were identified as significant contributors to the spread and development of a number of diseases. One example of this intellectual shift can be clearly seen in relation to *lamsiekte*, which Theiler began to investigate in 1912 in response to pressure from farmers in the potentially productive cattle-rearing district of Griqualand West. The condition was characterised by varying degrees of paralysis, often followed by sudden death. Field experiments at Armoedsvlakte Farm, near Vryburg, dominated research in the 1910s and revealed that *lamsiekte* was a form of botulism, acquired by gnawing carrion scattered around the veld. Cattle chewed the bones because of a condition known as pica, caused by a phosphorous deficiency in the soil and hence in the vegetation.

Once Theiler and his team had identified a link between phosphorous deficiency and susceptibility to *lamsiekte*, pedological and dietary investigations became important areas of research and provided a new dimension to the role of the veterinary scientist. For Petrus du Toit his profession went beyond curing or preventing disease:

> The chief aim of this Division is to maintain the health of our livestock. We emphasise the word health, because we wish to dispel the idea that the Veterinary Division is concerned only with the diseases of animals. Disease is the negation of health; and our endeavour is to prevent disease and to assist our farmers to maintain their animals in good health and condition.

Veterinarians had recommended the use of bone-meal in phosphorous deficient areas from the late nineteenth century, but subsequent research showed its properties also helped increase milk yields, promote weight gain and enhance reproductive performance. From the 1920s, biochemists analysed other supplements, including minerals such as salt and sulphur, as well as organic products such as fishmeal.

Further veld research involved liaison with the Botany/Plant Pathology Department, headed by Illtyd Pole Evans, and in the 1930s with Heinrich du Toit's Extension Division. Veld investigations led to the decentralisation of research from Onderstepoort to various field stations, located in different vegetative environments throughout the country. One of the most important was Fauresmith in the Orange Free State. Here the Swiss botanist Marguerite Henrici, one of the few women employed by the state at this time, combined research into the nutritive qualities of Karoo veld with investigatins into poisonous plants and animal health.

Toxicology came to the fore in the 1930s and combined ecological with chemical disciplines in an attempt to understand the cause of several important non-infectious diseases. In the past, Theiler and others had carried out drenching experiments, in which animals were force-fed plants suspected of being toxic. It was research at a crude level, based on observation of cause and effect, with the Botany Department identifying the noxious plant. Advances in technology and chemical processing meant that by the 1930s researchers could analyse material at the molecular rather than just the microscopic level.

Toxicologists managed to isolate the toxic principle ofvarious poisonous plants whilst physiologists looked at the anatomical effects of toxicosis. John Quin and Claude Rimington, for instance, explicated how *Tribulus terrestris* brought on the symptoms of *geeldikkop*. This was a particularly nasty disease involving severe liver damage, photosensitivity and facial disfigurement, and caused the death of thousands of sheep in the Karoo each year. Unfortunately these discoveries did not result in any effective cures due to the prohibitive cost of constant reapplications, the rapidity of death in some cases, as well as limitations in pharmacological knowledge. Identification of toxic plants did however feed into broader policies of weed eradication. William Beinart has discussed the government efforts in the 1930s to eradicate jointed cactus and the spiky variety of prickly pear (*doornblad*), but the state also invested mney in uprooting dangerous plants such as *gifblaar*.

Starting from the 1920s, fauna as well as flora became a focal point of research as scentists explored the relationship between game and the existence of enzootic diseases, such as nagana in Zululand. The bacteriologist, David Bruce, had proved a link between nagana, game and tsetse flies in the 1890s (discussed below), but a decline in incidence after the rinderpest epizootic (1896–7), which killed many cattle and faunal carriers, stalled further investigations for over twenty years. An increase in cases following the establishment of settlements, such as that at Ntambanana for ex-soldiers after the First World War, generated new enquiries, however.

The political construction of nagana was different from that pertaining to some other animal diseases, because it was directly linked to the future viability of white settlement. Veterinarians dealt with contagious and highly fatal infections, such as rinderpest and East Coast fever, as national catastrophes, and the rhetoric and procedures focused on saving the national, as opposed to just the settler, herd. In the case of nagana, however, incidence was due to envirnmental factors rather than direct contagion, and for the next three decades, the imediate catalyst for particular research initiatives was always the plight of white farmers. In 1925, Petrus du Toit specifically linked the development of Zululand as a 'white man's country' with the ability to deal with the fly.

Research into nagana recommenced in 1921 when Herbert Curson and Robert Harris from the Veterinary and Entomology Departments respectively, investigated the extent of trypanosome presence in game. Both found the parasites in the blood of various species and their findings came to have political, as well as veterinary, implications. Some entomologists and veterinarians became embroiled in acrimonious debates with conservationists over the existence of the Umfolozi-Hluhluwe and the Mkuzi Game reserves in Zululand and the founding of the Kruger Park in the Transvaal lowveld in 1926.

The entomologist, Claude Fuller, who investigated the disappearance of the tsetse species, *Glossina morsitans*, from the low veld following the 1896–97 rinderpest epizootic, was adamant that the creation of the Kruger Park would create a conduit of disease, enabling the reintroduction of this fly from Southern Rhodesia and Portuguese East Africa where it was still prevalent. Fuller had the support of leading veterinarians including Arnold Theiler and Petrus du Toit. Du Toit argued in respect of the Zululand reserves, that game should not be maintained in the proximity of human settlement.

The survival of all these game reserves, however, revealed political support for the conservationist cause and the belief that fauna should be protected as an economic and cultural asset. It also highlighted contradictions in government policy over land use and illustrated some limits to the political influence of South Africa's scientific elite even when wildlife threatened the profits of white settlers.

Nagana research continued into the 1930s alongside a revival in immunological investigations, brought about by international developments in vaccine manufacture, rather than the arrival of an unfamiliar and dangerous epizootic. Of particular significance was Raymond Alexander's invention of an inoculation against horse sickness in 1934.

Alexander copied a technique pioneered by Max Theiler (son of Arnold) who had devised a yellow fever vaccine, attenuating the virus by passage through the brains of mice. Theiler worked at the Rockefeller Institute in the United States, indicating the importance of personal connections in the creation of scientific networks and the growing influence of American-led biomedical research. Alexander's neurotropic vaccine was effective and opened up new areas of the country to horse breeding, namely in the Transvaal lowveld and the littoral regions of Natal and the Eastern Cape. Other technical innovations from abroad, in particular the invention of pesticides such as DDT by Swiss scientists during the Second World War, provided new opportunities for disease control. In a bid to intensify cattle production in Zululand, the veterinary entomologist, René du Toit, worked with the South African air force and perfected methods of disseminating DDT as a gas through the exhausts of planes, thereby eradicating the primary tsetse species, *Glossina pallidipes*, from the region by 1953.

Technological developments emanating both from within South Africa and outside were therefore important determinants in identifying why certain aspects of research or disease control were adopted at particular points in time. To this can be added pressure from farmers themselves; for example, in relation to the *lamsiekte* investigations. Although South African politics placed economic emphasis on the expansion of settler agriculture, Africans, too, benefited from veterinary developments. Evidently the health of settler livestock depended on disease eradication amongst animals kept by African tenants, as well as in the neighbouring African reserves. Even so, Petrus du Toit also argued that the benefits of veterinary medicine had to be made available and affordable to all, and that Africans, as well as settlers, should have the opporunity to make a viable living from the land.

Nevertheless, at times there was a clear differentiation in the allocation of veterinary services, as the timing of the nagana research demonstrated. But this was not always the case and Africans were not necessarily excluded from access to biomedical advances. In the case of East Coast fever, for instance, the Transkei became a field laboratory in 1912 for testing out a vaccine designed in Onderstepoort, and over 159,000 animals were inoculated during that year.

In parts of the Transvaal and Natal, East Coast fever had already resulted in a 95 per cent mortality rate amongst some herds, whilst this inoculation, although badly flawed, could save approximately 35 per cent of bovines exposed to infection. On the one hand this showed that veterinary researchers were prepared to take risks with African cattle, which they were less willing to take with settler herds due to potential compensation claims. On the other hand, however, when faced with the reality of the epizootic, many Africans wanted the vaccine i the hope of saving at least some of their animals, as a result of which the Veterinary Department was unable to cope with demand. On this occasion it was Africans who benfited, however imperfectly, from new technology. Farmers in other parts of the country were not provided with this vaccine and, by law at least, were banned from experimenting with inoculations lest this set up new centres of infection. The willingness of some Transkei pastoralists to accept this procedure, despite the risks, showed a certain receptiveness towards vaccination, which can be contrasted with the sometimes bitter encounters between cattle owners and stock inspectors, responsible for enforcing veterinary regulations.

Many farmers, both African and settler, disliked regulations and wanted vaccines to tackle all dangerous livestock diseases because they saw them as the least intrusive and most effective form of prophylaxis. In 1908, there were only two bacterial and four viral inoculations on the market. By 1958, Onderstepoort manufactured ten bacterial, seven viral, two ricketssial and one protozoal vaccine. The number of doses issued in 1908 amounted to less than 400,000. In 1958, sales surpassed 55,000,000, suggesting that many farmers were using this treatment on their animals. Unfortunately, scientists were unable to design inoculations for many serious tick-borne diseases, as well as nagana, and Onderstepoort had to withdraw the disappointing East Coast fever vaccine. Immunology had therefore failed to provide a universal method of disease control and in the absence of biomedical solutions the state resorted to an array of veterinary directives, including obligatory stock dipping, local quarantines and restrictions on animal movements.

The state imposed these measures on African and settler communities alike, and complaints came from all sectors of rural society. Dipping, in particular, was unpopular because it was costly, time consuming and also represented an unwelcome state intervention into the management of livestock. In East Coast fever areas, where dipping was compulsory, veterinarians recommended immersing stock every three to five days to combat the brown tick. In the Transkei this provoked localised unrest in 1914, and Africans, not only in the Eastern Cape, remained divided over the issue.

In 1923, for instance, some magistrates in Zululand reported that many people believed dipping had saved animals in a region subject to both East Coast fever and redwater. Others, however, emphasised African hostility, primarily because the inability to move cattle without cleansing impeded transhumance, stock sales, transport riding and the transfer of *lobola* (bridewealth) payments, all of which exacerbated poverty. Herding cattle to communal dipping tanks was also especially onerous for women whose husbands were absent at the mines. The imposition of veterinary regulations, in the wake of extensive male migrancy, therefore also had important social impacts and began to erode the traditional taboos against females handling cattle whilst increasing the labour burden designated to women.

It was not until 1955 that Onderstepoort could declare South Africa free of East Coast fever. By this time mortality from many other diseases had also declined and stock numbers had grown. Between 1911 and 1955, cattle numbers doubled from 5,796,949 to 11,689,475, whilst those of sheep rose from 30,656,659 to 37,042,504. Yet the benefits of veterinary medicine were not without their contradictions, as scientific advances could potentially reduce rather than enhance the productivity of the livestock industry. Disease control facilitated overstocking and thus overgrazing, thereby giving rise to erosion and the propagation of more resilient, but often less nutritious and potentially more toxic, wild plants. Optimising livestock yields was predicated not just upon biomedicine but also upon a more ecologically conscious approach to veld management.

Veterinarians were often critical of overstocking by both African communities and settler farmers. The toxicologist, Douw Steyn, for instance, advised farmers that overgrazing was the principal reason for the increase in mortality through plant poisoning. With the suppression of epizootics, plant poisoning, together with wirewom infestation, led to more fatalities than any other cause. Along with their colleagues in the Botany and Extension Divisions, veterinarians saw rotational grazing, in conjunction with fodder cultivation and the digging of boreholes to supply fresh as opposed to stagnant water, as vital for the development of a sound pastoral economy. Veterinary science alone could not achieve that end.

Veterinary Research

he expansion of veterinary medicine was an important illustration of how applied science and technology could enhance pastoral output. At the same time, it also demonstrated a cultural value attributed to science as a means by which South Africa could promote itself as a modernising state during the first part of the twentieth century. The link between colonial science and national identity has attracted the attention of a number of historians exploring the evolution of scientific institutions, networks and practices beyond the metropole. Jan Todd, for instance, has used the development of the anthrax vaccine to illustrate how local farmers, as well as scientists, have played a key role in adapting European science to suit a colonial, in this case the Australian, disease environment. Others, such as Saul Dubow, have written about the indigenisation of knowledge systems and suggested that the creation of a professional scientific elite was an important component in the construction of a white Anglo-Afrikaner 'South Africanist' identity. These approaches, together with Roy MacLeod's concept of a 'moving metropolis', help inform some aspects of the development of veterinary science in South Africa.

This section of the article explores how Onderstepoort epitomised the transferral and modification of western learning to meet the demands of the South African colonial periphery. It also illustrates the significance of local factors, in particular the voice of farmers and the nature of the environment, in the expansion of South African and, as a corollary, international veterinary knowledge. Some scientists also attributed a cultural value to South Africa's disease landscape that formed part of a constructed national identity.

The introduction of western biomedicine to South Africa demonstrated the failure of folk or indigenous medicine, often based on herbal concoctions laced with caffeine and Cape *dop*, to eliminate disease and promote sufficient growth in the livestock population, at least in the eyes of the capitalist and progressively minded rural elite. In the Cape and Natal, farmers successfully lobbied the state for the appointment of European-trained veterinarians in the 1870s. For the next 50 years many of South Africa's veterinary scientists came from Europe, in particular Britain, Ireland, Switzerland and Germany. After Union, South Africans such as Petrus du Toit, Philip Viljoen and Douw Steyn went to Europe to study. Berlin was particularly popular for tropical medicine; Vienna for toxicology. Continued links with the metropole ensured the flow of European (as opposed to simply British) science to the southern hemisphere.

Practitioners such as Arnold Theiler, however, were not happy with this state of affairs. Theiler demanded the indigenisation of veterinary education through the founding of a university faculty at Onderstepoort. He clearly revealed his motives for this in remarks made about an anonymous visiting researcher at Onderstepoort in the 1910s, when,

I realised once again how much the oversea [sic] veterinarian is out of touch with our problems, and although they may have a thorough grip of veterinary science as applicable to their own country, it sometimes seems to me almost hopeless when I realise how little they know about our own conditions. There is no doubt about it that we have our own veterinary science, and in this respect South Africa is undoubtedly unique.

In 1920 the government established Africa's first veterinary faculty, affiliated to the University of Pretoria. This once again emphasised the importance of the livestock industry to the South African economy and the influence veterinary scientists could have over state policy. It also resonated with a process that in 1929 Jan Hofmeyr termed the 'South Africanisation' of science. South Africans, as opposed to foreign experts, were now working on local problems with the result that 'science in South Africa. . .has made itself truly South African'.

The first cohort of South African veterinary students arrived at Onderstepoort in 1921. Given that much of the medical literature at the time was written in French and German, students had to be proficient in one or both of those languages, as well as fluent in English and Afrikaans. The course lasted five years and included a wide range of veterinary subjects, as well as topics more specifically suited to South African conditions, such as parasitology, which dealt with worms (helminths) and external parasites such as ticks. Gradually researchers at Onderstepoort began to publish standard textbooks reflecting the localisation of veterinary literature and the declining dependence on publications from overseas. Hermann Mönnig compiled South Africa's first book on helminthology in 1934, whilst Douw Steyn's work o poisonous plants remained the principal point of reference for toxicology until the 1980s.

Reflecting the gendered and racial divisions of labour of the day, most of the people who trained and researched at Onderstepoort were white males. By the late 1940s however, women began to appear, albeit in minuscule numbers. In 1948, three out of the 61 researchers on the Onderstepoort staff list were female. Africans worked at the laboratories and participated in field experiments as assistants. There was no move to train African veterinarians, although some of the research staff did not appear to have been averse to that idea. In 1926, the Transkei Territories Native Council asked Theiler whether he was prepared to admit black students. Theiler to some extent avoided the question by stating that that was a political issue but from his personal perspective, he did 'not see how the native can be rightfully prevented from qualifying'. Several years later, when field staff were questioned about what could be done to improve the service, J.J. Keppel, working in Cape Town, argued that the state had to train African veterinarians and provide them with the opportunity to obtain degrees so that they could practise in the reserves. Whether Keppel believed African veterinarians should be excluded from treating settler livestock was not made clear, but his comments problematised the racial agenda by indicating that some scientists did not regard their discipline as the preserve of a white male elite and believed that the indigenisation of the veterinary service should include Africans, too.

The 'South Africanisation' of science involved more than the training of experts. It also entailed interaction with farming communities. Farmers and herders had first-hand experience of the environmental factors that undermined livestock rearing. They could therefore contribute ideas about possible lines of research whilst their agricultural methods provided models that could be scientifically assessed as poor or sustainable husbandry practices. Many diseases, especially those conveyed by arthropods or those that emanated from deficiencies in the veld, were not the same as those found in Europe, thereby enabling farmers to play a pivotal role in the expansion of South Africa's veterinary knowledge. Khoisan and Xhosa herders, for instance, postulated a link between the '*nenta*' plant and *krimpsiekte* in goats. Farmers also provided vital information regarding the distribution of ticks and brought to entomological attention new species for identification. Entomologists believed that an understanding of the anatomy, the lifecycle and the habitat of arthropods was fundamental to their control or eradication. Literate and enterprising farmers, in turn, used this knowledge to devise their own practical means of combating these pests. In the first decade of the twentieth century, Joseph Baynes from Nels Rust in Natal, worked with the bacteriologist, Herbert Watkins-Pitchford, and put Lounsbury's analysis of the lifecycle of the brown tick to practical use by producing a three-day arsenical dip to cleanse his stock. Veterinarians later introduced this method to East Coast fever areas. The expansion of veterinary knowledge and the refinement of methods of disease control therefore involved a two-way interaction between scientists and farmers.

African observations of the natural environment and the way that it could be manipulated were also important in relation to the way South African veterinarians, as well as other colonial authorities, would deal with nagana. Statements by officials and European travellers indicated that the Zulu had a clear understanding of the disease, which pre-dated David Bruce's reports first publishedin 1895. In 1891, the

Resident Commissioner of Natal, F. Cardew reported that: 'The natives assert that the disease is caused by cattle when grazing where big game abound eating the saliva of the latter left on the herbage'. The British traveller, E.C. Buxton, also noted that '[t]he belief of the natives in the dangerous character of the fly is universal'. When Bruce detected trypanosomes in animals ailing from nagana, as well as in the blood of various species of game, he provided a bacteriological corroboration of Zulu ideas about a faunal link and he also proved that tsetse flies transmitted the disease. Bruce's reports demonstrated an interrelationship between indigenous knowledge emanating from the colonial periphery and founded on close observation of the local environment and western biomedical systems sponsored by the European metropole: in this instance each informed and consolidated the other.

Almost 30 years later, Herbert Curson commented positively on how the Zulu used their understanding of the cause of nagana to protect their cattle from tsetse flies by periodic game slaughter, veld burning and by cultivating along river banks, thereby removing the dense thickets favoured by *Glossina pallidipes*. Settler farmers, however, failed to clear the bush properly and they overstocked their farms, giving rise to denser secondary vegetation, which forced animals to graze on more marginal and fly-infested grasses, especially in dry seasons.

Curson's report indicated a sensitivity to and appreciation of African methods of farming, which is particularly interesting because much of the historical analysis has focused on those cases where scientists criticised African farming as wasteful and environmentally destructive. For Curson, however, whites, not blacks, were responsible for propagating disease by misusing the environment.

With colonialism, however, Africans lost some of their ability to manage their surroundings and tackle diseases such as nagana on their own terms. In particular, the introduction of restrictions on hunting, as well as the founding of game reserves in Natal, during the 1890s prevented Africans from removing wildlife at will. As already discussed, however, the presence of nagana meant conservation was a problematic issue in Zululand and, despite opposition from the wildlife lobbies, the Veterinary Department was able to copy Zulu precedents by organising game drives and slaughter campaigns periodically from the 1920s onwards.

By the 1940s, scientists also tacitly endorsed Zulu methods of bush clearance as they used bulldozers and chains to clear vegetation from identified tsetse breeding sites, riverbanks and pathways frequented by game. Researchers at Onderstepoort therefore learnt much from African experiences, which were especially important in dealing with diseases for which there were no vaccines. In the case of nagana, scientists introduced modern technology in the form of guns and machinery to clear game and vegetation, but the underlying reasons for doing so originated with the Zulu.

In contrast to the immediacy of disease control lay the long-term problem of conserving the veld and promoting animal health. The natural veld symbolised the South African landscape and from the 1920s onwards the Veterinary Department collaborated with botanists and ecologists in the Plant Industries Division, as well as with universities, and in particular the University of the Witwatersrand. Granting primacy to the indigenous grasslands represented a change in environmental thinking since the late nineteenth and early twentieth centuries when scientists had made assumptions that South African trees and vegetation were deficient compared with the trees and vegetation of other continents. Now the emphasis was on the fact that the South African environment was harsh – subject to frequent droughts, violent storms, easily eroded soils and fragile grasses – but not that it was intrinsically inferior.

It simply required careful management. Some scientists took great pride in the assumption that the South African environment was somehow unique. Douw Steyn, for example, revelled in the fact that South Africa had a plethora ofpoisonous plants, which were part of the country's rural identity and provided exciting opportunities for research. Similarly, Theiler boldly stated that '*Gousiekte* is a disease of South Africa', whilst the biochemist A.I. Malan argued for a South Africanist approach to nutritional research: 'The problems of animal nutrition must be considered as national problems and we must not be content to look to other countries for help and solutions'.

Work on the veld showed that South African scientists had embraced the new science of ecology. In this context debates about overgrazing became more involved. Although much of the rhetoric still contained a critique of overstocking and kraaling, some of the arguments were not so simple. Each animal species, whether wild or domestic, engaged in different grazing habits that were now analysed within the remit of the ecological language of vegetation succession and climax. Botanists and toxicologists discussed popular perceptions that plant poisoning was becoming more common in terms of an increase in secondary or *opslag* grasses that were less palatable and possibly poisonous to stock. Veld research and grazing practices became increasingly localised, because flora could be poisonous on one farm and not on another due to different soils that affected the chemical composition of plants. Paddock rotation, therefore, came to mean more than just resting sections of the grasslands to enable regeneration. It also meant having the right proportion of different animals on the land, which had to be worked out on a farm-by-farm basis. Philip Viljoen, amongst others, invoked the language of a 'balance of nature' and suggested that it was essential to restore or create a 'natural' equilibrium through sophisticated grazing practices, based on a careful assessment of the carrying capacity of the land.

Viljoen also emphasised the importance of indigenous, as opposed to exotic, grasses, along with the need to cultivate fodder. The indigenisation of grassland research reflected a dramatic shift from the acclimatisation experiments using alien flora that had characterised both veld and forest policies in the early twentieth century. Ecology encouraged a local focus, and international ideas about plant succession became 'South Africanised' through their specific application to the veld.

Moving from the very local environment to a broader geographical field, the 'South Africanisation' of the country's scientific culture and practice contributed to the emergence of a sub-metropolitan science in which South Africa, to some extent, replaced Europe as the font of veterinary authority in parts of Africa. Some colonial veterinarians had consulted Theiler and others prior to the founding of Onderstepoort, but, from the 1920s onwards, this *rapprochement* became far more commonplace and it was South African nationals, as opposed to foreign-trained scientists, who provided this expertise. Tony Kirk-Greene has noted that the size of the British colonial veterinary service in Africa was particularly small.

This provided opportunities for practitioners from Onderstepoort, and Britain affirmed its belief in its skills by inviting South African veterinarians to investigate disease in other parts of the Empire, such as Nigeria (toured by Petrus du Toit in 1927). South Africa's success in eliminating rinderpest, East Coast fever and nagana boosted institutional self-confidence and encouraged scientists at Onderstepoort to take the lead in African disease eradication campaigns. During the Second World War, for example, D.T. Mitchell set up a mobile laboratory on the Tanganyikan plains to produce vaccines against rinderpest, whilst René du Toit and others later advised the Tanganyikan and the Southern Rhodesian governments on the destruction of tsetse belts. Having cleared the country of epizootics that had hampered pastoral development in the past, South Africa was determined to mitigate the possibilities of reinfection from the north. In turn, other African states wished to benefit from South African knowledge and experiences, as these were often more appropriate for the continent's disease environment than discoveries made in the temperate European metropole.

Conclusion

Onderstepoort Veterinary Institute made an important contribution to South Africa's economic development as well as to its scientific culture. The Institute was born out of an imperial and South African material ideal of agricultural progress. Increasingly, veterinary medicine developed as an inclusive rather than a discrete science, and by the 1920s a more holistic approach to animal health, as explicated by Petrus du Toit, embraced the ecological, toxicological and biochemical disciplines, as well as the purely biomedical.

Nevertheless, veterinary science involved more than just cures and prophylaxis, accompanied by strategic grazing and supplementary feeds; it also encompassed more coercive dimensions. Governments used scientific knowledge to legitimise greater state intervention into the lives of rural communities, and at times veterinary policing superseded animal husbandry as a method of disease control. This can be seen most clearly with respect to East Coast fever, which took 50 years to eradicate. During that time many people suffered considerable economic hardship due to the cost of dipping and restrictions on stock movements. In the longer term, however, eradication of this disease, along with *Glossina pallidipes*, did provide new opportunities for cattle farming.

This was complemented by a large increase in sheep and cattle numbers despite the presence of numerous infections and recurrent periods of severe drought. Of course, it is impossible to determine conclusively the extent to which veterinary science alone contributed to this increase in this process, given that other variables such as political and economic contingencies, as well as weather patterns, available grazing and changing farming practices, affected livestock populations too. Nonetheless, the fact that much of this growth occurred during the first three decades of the twentieth century, corresponding to the decline in epizootics, implied that the development of vaccines and the enforcement of quarantines significantly influenced this outcome. From the 1930s, the livestock population began to stabilise as epizootics gave way to poisonous plants and internal parasites as the major killers. This in itself demonstrated contradictions in the beneficence of veterinary science. Scientists attributed the growing danger of toxicosis to overstocking, which biomedical advances had helped to promote. The expansion of veterinary research also reflected a cultural belief in science as a tool for ameliorating the natural world.

Veterinarians regarded scientific applications, as well as legislative interventions, as essential features of a modernising state that strove towards self-sufficiency not only in terms of pastoral output but also n relation to the country's ability to resolve its own epidemiological and environmental problems. Arnold Theiler was instruental in pushing for the establishment of a university faculty at Onderstepoort and 'South Africanising' veterinary medicine in order to end the country's dependence on European scientists and their publications. The consolidation of a national body of veterinary knowledge came to be rooted partly in metropolitan biomedical systems and partly in the scientific analysis of the observations and methods used by local farmers.

Africans in particular made an important contribution to the identification of toxic plants as well as to the aetiology and control of nagana. Comparative success in tackling a number of dangerous stock diseases boosted institutional self-confidence and encouraged South African scientists to intervene in disease control campaigns beyond South Africa's borders.

Despite the significant role played by Africans in the expansion of veterinary knowledge, however, veterinary scientists tended to regard settler farming as the key to economic progress. In this respect, some historians have emphasised the exclusive advantages agricultural science brought to white commercial farmers. The priority attributed to settler interests was clearly shown in relation to veterinary interventions against nagana in Zululand from the 1920s onwards.

Nevertheless, this inequality in veterinary provision was not always so blatant, especially when it came to dealing with contagious diseases such as rinderpest, in which case vaccines were distributed to all. The East Coast fever inoculation campaign in the Transkei also complicated arguments about the provision of biomedical resources.

In that case, African cattle were the subjects of experiments but, at the same time, their owners had access to a procedure that was officially denied to others. Ultimately both African and settler farmers experienced an increase in their livestock holdings during the first half of the twentieth century, but the broader racially defined political and economic agendas ensured that the former could not benefit to the same degree as the latter from the material gains that veterinary science could bring.

Veterinary Treatment for Disease

Research has indicated that a number of different factors affect whether an animal receives treatment or not when diseased. The aim of this paper was to evaluate if herd or individual animal characteristics influence whether cattle receives veterinary treatment for disease, and thereby also introduce misclassification in the disease recording system.

Methods

The data consisted mainly of disease events reported by farmers during 2004. We modelled odds of receiving veterinary treatment when diseased, using two-level logistic regression models for cows and young animals (calves and heifers), respectively. Model parameters were estimated using three procedures, because these procedures have been shown, under some conditions, to produce biased estimates for multi-level models with binary outcomes.

Results

Cows located in herds mainly consisting of Swedish Holstein cows had higher odds for veterinary treatment than cows in herds mainly consisting of Swedish Red cows. Cows with a disease event early in lactation had higher odds for treatment than when the event occurred later in lactation. There were also higher odds for veterinary treatment of events for cows in January and April than in July and October. The odds for veterinary treatment of events in young animals were higher if the farmer appeared to be good at keeping records. Having a disease event at the same date as another animal increased the odds for veterinary treatment for all events in young animals, and for lameness, metabolic, udder and other disorders, but not for peripartum disorders, in cows. There were also differences in the odds for veterinary treatment between disease complexes, both for cows and young animals.

The random effect of herd was significant in both models and accounted for 40–44% of the variation in the cow model and 30–46% in the young animal model.

We conclude that cow and herd characteristics influence the odds for veterinary treatment and that this might bias the results from studies using data from the cattle disease database based on veterinary practice records.

Background

Previous research has indicated that a number of different factors may influence the farmers' treatment decisions for diseased dairy cattle, and whether a veterinarian is contacted or not. Nyman et al. noted that the threshold for contacting the veterinarian differed between dairy farmers in Sweden. Moreover, Vaarst et al. found that the decision about veterinary treatment in Danish dairy cows depended not only on the disease event's severity, but also on the age of the cow, lactational stage, milk yield and/or the temperament of the cow, and that the farmers weighted these factors differently. The economic value of an animal is also likely to affect the decision about veterinary treatment.

The individual dairy cows' retention pay-off values differ depending on, for example, their parity, stage of lactation and milk yield. Another example is the report by Ortman and Svensson where they found a high proportion of treatments in young animals initiated by the farmers themselves.

How farmers' decisions about treatments can influence data quality is exemplified by Mulder and colleagues who compared cows with complete data records (defined as "not having missing data for postpartum evaluation, pregnancy diagnosis and body condition score...") versus missing data records and found a lower reproductive performance in cows with complete data records. One possible explanation was that problem-cows were identified early, and treated more intensively.

These findings indicate that cow, herd and/or farmer characteristics may affect whether an animal receives treatment or not when diseased. As a consequence of this, misclassification of disease events in animal disease recording systems based on veterinary treatments could be differential, i.e. occur with different magnitudes and with different directions. Note that the source of misclassification in this report is data loss – because animals are classified as healthy when there are no records saying they are diseased. Secondary databases with disease information have been used for research in several scientific areas, such as epidemiology, genetics and animal health economics. The advantage of such secondary databases is the large amount of data available at a low cost. However, a disadvantage is that the researcher does not have control over the data collection and consequently not of the data quality either. To ensure the data quality, a secondary database needs to be validated.

The dairy disease database (DDD) at the Swedish Dairy Association is based primarily on clinical disease events reported by veterinarians and is used for sire evaluation, extension services, annual statistics and research. The DDD has been evaluated concerning completeness with respect to all the disease events observed by farmers and also in respect of disease events resulting in veterinary treatment (Mörk et al., unpublished). It was found that only 54% of the disease events detected by farmers were treated by veterinarians. Consequently, the incidence rates for different disease complexes, based on the reported events, were significantly lower compared to the incidence rates based on farmer observations. It would be of interest to characterise this loss of data by examining if any animal and herd factors influenced whether a diseased animal received veterinary treatment or not.

Hence, the objective of this study was to evaluate if herd or individual animal characteristics influence whether a cow or young animal receives veterinary treatment for disease, and thereby also introduce differential misclassification in the disease recording system.

Materials and Methods

The study population and data collection have been described previously. In brief, a baseline study of disease incidence in dairy farms, based on farmers' records, was performed during January, April, July and October in 2004. Four-hundred herds were randomly selected from all the herds enrolled in the Swedish Official Milk Recording Scheme, which included about 86% of the Swedish dairy cows in 2004, and 177 participated.

The farmers were asked to record disease events, defined as an observed deviation in health. The farmer could either choose to wait, treat the animals him/herself, contact a veterinarian for diagnosis and treatment, or slaughter the animal. The data reported for each disease event were as follows: the animal's identity and sex, the date when the disease was observed, the diagnosis, whether or not a veterinarian was consulted, the farmer's description of the event (e.g. symptoms) and the treatment given. The farmers could use the following diagnoses: acetonemia/inappetence, abomasal displacement, calving problems, clinical mastitis, clinical puerperal paresis, coughing, diarrhoea, lameness (of a hoof), lameness (of a limb), retained placenta and other diseases. During data editing the diagnosis "other disease" was categorised into gastro-intestinal disorders, laminitis, paresis (not puerperal), peripartum disorders (retained placenta and puerperal paresis not included), ringworm/lice, traumatic reticuloperitonitis, udder disorders and other disorders based on the descriptions provided by the farmer. The farmer did not have to report the animals' identities for events where groups of animals were affected.

When comparing the disease events reported by the farmers with those reported by veterinarians in the DDD, it could be observed that some events resulting in veterinary treatment had not been reported to us by the farmers. Such events were also included in the analyses in this study.

Data from the Swedish Official Milk Recording Scheme

Information about the herds studied was obtained from the Milk Recording Scheme at the Swedish Dairy Association in November 2005. The data consisted of herd characteristics, such as the housing system and the herd size, as well as individual cow parameters, such as parity, calving dates, milk yield, fertility treatments and disease events.

Data Editing

The herds were categorised as Swedish Red (SR) herds or Swedish Holstein (SH) herds if at least 80% of the animals were pure-bred SR or SH, respectively, and as mixed/other breeds otherwise. Further herd-level characteristics were: the average milk yield in 2003 (calculated as the total daily milk yield in the herd/total number of cow-days for lactating cows), the average parity, the proportion of older cows (above the 2nd and 3rd lactation, respectively), the average somatic-cell count (SCC) in test milk, the average udder-disease score (UDS) and the proportion of cows with a high UDS, indicative of sub-clinical mastitis. The UDS is used to measure the probability of a cow having mastitis and is based upon a series of three test day SCC results at monthly intervals for individual cow's SCC. The variables were checked for implausible values, but none were found.

The data on animals with disease events reported by the farmers were merged with data from the milk recording scheme and the DDD and categorised into young animals (prior to the first calving for heifers) and cows, respectively. All the bulls with a reported disease event were below 2 months of age. The following cow characteristics were available: the breed, the parity, the milk yield on the test day prior to the disease event, the average SCC and average UDS, for the past 305-day period and 90-day period respectively, the days in milk at the disease event, the state of pregnancy at the time of the disease event (yes/no) and the number of in seminations prior to previous and current pregnancy.

Information was also available as to whether the cow was culled or not after the disease event (not culled, culled within lactation, culled after the current lactation) and the reported reason for culling. For young animals, the age at the time of the disease and the breed were available in the data from the Milk Recording Scheme.

In the current study, for cows, the disorders were categorised into the following disease complexes: lameness, metabolic, peripartum disorders (puerperal paresis not included), udder disorders and other disorders; and for young animals: coughing, diarrhoea, lameness and other disorders. Only four young animals had udder disorders, and these were categorised as other disorders.

In a previous study we found that the study farmers reported only 88% of the disease events reported to the DDD (by veterinarians) to us. Based on that finding, the farmers with an apparently good record-keeping ability were identified. Farmers qualifying as good record keepers had accomplished one of the following: i) they had reported all the events registered in the DDD (by veterinarians), or ii) they had failed to report one event registered in the DDD, but the number of successfully reported events was > 1, or iii) they had failed to report e” 2 events registered in the DDD, but the proportion of successfully reported events was > 0.75. The reasoning behind the different definition for farmers with one and more than one event missing was that one event was thought to be easily forgotten without necessarily indicating a waning interest in the study.

Missing Data

Fifteen animals were dropped because data on them were missing. Moreover, for 34 cows with incomplete calving data, an approximate calving date was calculated for estimation of the milk yield in the latest 305 day-period. This was accomplished by subtracting the study population's median calving interval (384 days) from the following calving date.

Statistical Analysis

Two-level regression models were fitted with a logit-link. The dependent variable was whether the diagnostic event had resulted in veterinary treatment (yes = 1/no = 0), as reported by the farmers (or the DDD). Veterinary treated refers to all events where the farmer contacted a veterinarian, even if no medical treatment was delivered. The probability ($p^{\text{veterinary treatment}}$) of veterinary treatment for an animal with a diagnostic event depended on explanatory variables ($x_1,..., x_n$) and on the random effect of herd (u_{herd}). The random effect was assumed to be independently and normally distributed with a standard deviation, $ó_u$. The model for animal i was expressed (using logit $(p) = \log (p/(1 - p)))$ as the data on cows and young animals were analysed separately in two models. Only events where the animal's identity number was reported were used in the analysis (i.e. no events reported for groups. The group reported events were: eight and nine events of herd-outbreaks of cough and diarrhoea, respectively). Further, 326 cows had more than one diagnostic event, either at the same date or at different dates. Only one diagnostic event per cow was included in the analysis to avoid the effects of clustering on the individual animal level. These events were selected by giving the events a random number and including the one with the lowest number. Moreover, we included in our models only disease events from herds with at least four disease events in dairy cows or young animals, respectively.

The continuous variables, except the herd's average milk yield, were not linearly related to the outcome (based on logit-transformed smoothed scatterplots) and therefore categorised using the 25th, 50th, and 75th percentiles. The association between the outcome and each potential fixed explanatory variable was tested in a univariable analysis including herd as a random effect. By including a dispersion parameter in the empty two-level models, we estimated the extra-binomial variation to be 0.86 for cows and 0.94 for young animals. Since it was reasonably close to 1, the dispersion parameter was not considered in further analyses.

The final models were, however, re-fitted with the dispersion parameter included, resulting in no changes in the estimates in the cow model and only small (less than 0.1) changes in the estimates in the young animal model and only results from the models without the dispersion parameter is presented. Correlations between the explanatory variables considered for further analysis were investigated using Spearman correlation coefficients, with the intention of dropping one of the variables if the correlation was e" 0.7 or d" -0.7. In the analyses for cows, the herd's proportion of cows above the third lactation and the herd's average parity had a correlation coefficient of 0.8 and the herd's average parity was therefore excluded in the multivariable analysis. In the analyses for young animals, no variables were dropped.

All the explanatory variables with a p-value < 0.2 (in the likelihood ratio test) in the univariable analyses and no missing observations were included in the multivariable analysis. The model was reduced manually by backward elimination. A variable with a p-value d" 0.05 (in the likelihood ratio test) was considered statistically significant and kept in the final model. All the variables excluded were then re-entered, one at a time, and kept if their p-value was d" 0.05. All the two-way interactions were then tested for inclusion one by one. A variable was considered to be a confounder, and therefore retained in the model regardless of significance tests, if deleting it from the model resulted in the change of another parameter estimate by more than 20%. The variance partition coefficient (VPC) was estimated by (ó$^{2}_{\text{herd-level}}$/ó$^{2}_{\text{herd-level}}$ + ó$^{2}_{\text{event-level}}$), where we assumed that the level-one (event) variance was ð^{2}/3 (where ð = 3.1416) on the logit scale.

Data editing, descriptive statistics and model building (log likelihood estimation (LL) using the xtmelogit command) were performed in Stata® version 10 (Stata Corporation, College Station, TX, USA). The final models were also estimated using the second-order penalised quasi-likelihood (PQL) and the restricted iterative generalised square algorithm and the Markov-chain Monte Carlo (MCMC) procedures in MLwiN (version 2.1, Institute of Education, University of London, UK). The evaluation of extra binomial variation was performed using the PQL estimation.

The MCMC model was fitted using the Metropolis-Hastings algorithm with diffuse priors, a burn-in length of 500 iterations and a monitoring period of 90,000 iterations. The model fit was evaluated by plotting the standardised residuals against the fixed part prediction and normal scores, respectively, at the second level (herd) for the PQL estimation. For the cow model and the young animal model, the points in the plot of standardised residuals against the fixed part prediction showed an equal-width band and the plot of standardised residuals against normal scores showed a, roughly, straight line. For the cow model, two possible outliers (standardised residual below -3) were detected but the model did not change much when those observations were deleted.

Results

In the original data for cows there were 2,112 diagnostic events in 171 herds. From these data, 338 events were deleted because of multiple events per cow. Moreover, 67 events in 31 herds were deleted because the herds had less than four events. The original young animal data contained 362 diagnostic events in 96 herds, of which 13 diagnostic events were deleted because of multiple events in one animal. Another 106 events and 68 herds were removed because the herds had less than four events. The resulting datasets consisted of 1,707 diagnostic events (in 140 herds) in cows and 243 diagnostic events (in 28 herds) in young animals.

For cows the average number of events per herd was 12.2 (with the median being 10, the range 4–88). Of all the events, the proportion that resulted in veterinary treatment per herd ranged between 0% and 100%, with the 10th, 50th and 90th percentiles being 21%, 75% and 100%. For young animals the average number of events per herd was 8.7 (with the median being 6, the range 4–37). The percentage of events resulting in veterinary treatment per herd ranged between 0% and 100%, with the 10th, 50th and 90th percentiles being 0%, 18% and 100%, respectively.

Logistic Regression Analysis

Cows

For cows, the categorical variables included in the multivariable analysis are presented. Variables with a p-value > 0.2 in the initial analysis, and thus not included were: good record-keeping ability, herd average SCC, herd size, parity, private or state-employed veterinary district, and the proportion of cows older than the second lactation.

The only continuous variable included in the multivariable analysis was the herd's average milk yield. The 10th, 50th and 90th percentiles for the herd's average milk yield (kg of energy-corrected milk) were 6,564, 7,903 and 9,344 for herds for which events resulting in veterinary treatment had been reported, and 6,313, 7,751 and 9,303 for herds for which events resulting in veterinary treatment had not been reported. The average milk yields per cow in the latest 305-day and 90-day periods had a p-value < 0.2 in the univariable analysis, but were not included in the multivariable analysis because of missing observations. Instead, cow average milk yield in the latest 305-day and 90-day periods were tested in a model containing the explanatory variables that remained in the final model. They were, however, not statistically significant.

One herd-level variable and four event-level variables were retained in the final model for cows. Moreover, the final model included an interaction between the disease complex and another animal with an event at the same date. The estimates and standard errors based on the LL, PQL and MCMC procedures were similar (Table).

The odds ratios for veterinary treatment for the LL estimation are presented in Table. The interaction term is presented as a comparison within the disease complex in Table Table3.3. The baseline for the interaction term was udder disorders combined with no other event at the same day. When no other animal in the herd had an event at the same date, lameness disorders had a statistically significantly lower odds for veterinary treatment than the other disease complexes (OR 0.37; 95% confidence interval (CI) 0.19, 0.70 compared to metabolic disorders; OR 0.35; 95% CI 0.16, 0.74 compared to other disorders; OR 0.28; 95% CI 0.12, 0.67 compared to peripartum disorders and OR 0.40; 95% CI 0.24, 0.67 compared to udder disorders). When another animal had an event at the same date, lameness disorders had a significantly lower odds for veterinary treatment than metabolic disorders (OR 0.19; 95% CI 0.08, 0.45), other disorders (OR 0.08; 95% CI 0.03, 0.28) and udder disorders (OR 0.09; 95% CI 0.05, 0.16), and peripartum disorders had a significantly lower OR than udder disorders (OR 0.17; 95% CI 0.06, 0.48) and other disorders (OR 0.15; 95% CI 0.04, 0.61).

Herd as a random factor was significant and accounted for 41% of the modelled variation in the LL estimation and 41% and 44% in the PQL and MCMC estimations, respectively.

Young Animals

The herd's average milk yield, the herd's average UDS, herd size, housing type and the proportion of cows older than the second and third lactation, respectively, were tested in the initial analysis for young animals, but had p-values > 0.2. The categorical candidate variables included in the multivariable analyses are presented in Table Table11.

The different estimation procedures showed different results for the young animals' analysis, with lower estimates for most variables in the LL procedure (Table (Table4).4). The ORs for the LL procedure are presented. Study month was identified as a confounder and was included in the final model although non-significant (p = 0.07, data not shown).

Moreover, the random herd effect varied between the LL, the PQL and the MCMC procedures and was significant in the LL and MCMC procedures, but not in the PQL procedure. The estimated variation ranged from 30–46%.

Discussion

This study deals with the probability of receiving veterinary treatment (i.e. the probability that the farmer contacted the veterinarian) for diseased animals. Thus, it is important to keep in mind that the explanatory variables significantly associated with the probability of receiving veterinary treatment are variables that seems to influence whether diseased animals receives veterinary treatment or not. Hence, they are not necessarily risk factors for disease.

Cows

Breed was the only statistically significant herd-level characteristic that affected the cow's probability of receiving veterinary treatment, with lower odds in SR breeds than in SH breeds. Several studies based on either farmer's disease records or veterinary records have found a difference in incidence of disease between breeds. This difference could have many possible explanations. The concentration of several blood variables have been found to differ around calving in primiparous cows of SH and SR breed and potentially explain why cows of SH-breed have higher disease incidence.

It could also be hypothesised that a difference in immune response between breeds could affect the severity of a disease event and thus the odds of receiving veterinary treatment. Nyman et al. found that herds with high incidence rates of clinical mastitis consisted more often of SH cows, and in these herds, the farmer more often contacted a veterinarian as soon as the cow's milk appearance was altered than in herds with low incidence rates. Persson Waller and her colleagues suggested that differences in treatment strategies between SR and SH herds could have biased the effect of breed in studies using veterinary records of disease, concurring with our findings.

The higher odds for veterinary treatment early in lactation could be expected, because the transition period (from three weeks before until three weeks after calving) is known to be associated with a higher risk of disease ; and especially as many diseases during this period have an acute course and demand veterinary assistance or treatment. Metabolic and physical stress related to pregnancy, parturition and lactation have been described to have a negative impact on the health. It is also possible that changes in the immune system during this time cause a more severe course of disease. Further, it is also possible that the lower odds for veterinary treatment later in lactation was affected by different treatment strategies for cows in different lactational stages, as has been shown for mastitis by Vaarst and her colleagues.

The interaction between diagnosis and whether or not there was another animal with an event at the same date resulted in the finding that there were higher odds for treatment of lameness, metabolic disorders, udder disorders and other disorders if there was another animal with an event at the same date. Animals with an event at the same date as another animal belonged to herds with a significantly higher herd size ($p < 0.001$, using the Wilcoxon rank-sum test, data not shown). It is likely that the veterinarian was consulted for milder disease events to a higher degree if he or she was contacted for another event. The cost for examining, and treating, a mild case will be lower when the veterinarian is already at the farm. A higher incidence of veterinary treatments in large herds could therefore be an effect of a higher number of events near in time in large herds, and not only an effect of a higher incidence of disease in larger herds. On the other hand, cows in smaller herd have a larger relative economic value and could therefore be more likely to receive veterinary treatment than cows in larger herds, as discussed by steras et al.. A difference in a variable at the individual level, i.e. the probability of veterinary treatment in the case of a disease event that relates to the group to which the individual belongs, is called a contextual effect. Contextual effects that are not accounted for could lead to false inferences, as shown by Stryhn and his colleagues. Herd size as well as herd main breed should therefore be regarded, and taken into account, as contextual effects.

Significant differences in odds for veterinary treatment between disease complexes were mainly found between lameness and the other disease complexes, with lameness having lower odds for veterinary treatment. For most events of lameness that were not veterinary treated, the farmer had reported that the hoof trimmer had been contacted. There is a voluntary hoof health register in Sweden, and a combination of the disease database and the hoof health register would give more complete information on hoof disorders. Reproductive disorders had significantly lower odds for veterinary treatment than udder disorders and other disorders in presence of another animal with a disease event at the same day. This could be seen as an indication of that those events of udder disorders and other disorders were milder and more likely to be consulted for only when the veterinarian was already at the farm.

Study month had an effect on the odds for veterinary treatment. Whether the animals are kept on pasture or housed indoors could affect the farmers' treatment strategy, as well as the ability to detect disease. Although our results could not reveal a seasonal pattern, a seasonal variation in the severity of the disease events and a seasonal difference in pathogen prevalence have been reported, for example for mastitis. A higher incidence of parturient paresis has also been found during grazing. It is, however, also possible that the differences between study months are an effect of our study design. January, the month when there is least to do at a farm under Swedish conditions, was the first study month, while October was the last. A lack of time during the harvest and a reduced interest in the study during the last months could have influenced the farmers' own recordings in favour of those events that resulted in veterinary treatment. Such events may be easier to recall, because the veterinarian leaves documentation on the farm after the treatment.

Consequences for Studies based on Veterinary Reported Data

This study has identified a number of factors that affects whether a diseased animal receives veterinary treatment or not. From the data it was not possible to fully distinguish severe disease event from those with milder symptoms of disease but it is likely that many of those not veterinary treated were milder disease events. It is also likely that for some events that were not veterinary treated the farmers decided to slaughter the animal instead of treating it. When using veterinary treated disease event in studies of risk factors for disease, the estimates of breed, lactational stage and likely also estimates of herd size could be biased. A number of variables were considered for inclusion in the models, but were not significantly associated with veterinary treatment such as housing types and milk yield (cow or herd average). Hence, based on our results, veterinary recording data could be used to study those risk factors without the risk of bias being introduced because of a differential veterinary treatment attributed to the risk factors being investigated.

Random Part

The original data consisted of events subclustered within individuals which in turn were clustered within herd. Because only

16% of the cows had more than one event, multiple events in animals were removed. The clustering of events within herds was accounted for by the random effect of herd which was significant in the cow model, with similar results from the different estimation procedures. In the model for young animals, the estimation procedures showed different results, and the random effect of herd was only significant for the LL and MCMC procedures. It is also possible that the random effect was over-estimated due to some herds having only events that resulted in veterinary treatment. Excluding these herds from the model reduced the herd-level variation to between 25% and 28% for cows and 24% and 39% for young animals (data not shown). It was, however, not possible to determine if the farmers whose herds only had events that had resulted in veterinary treatment had reported all the events or had failed to report events that had not resulted in veterinary treatment.

In the cow analysis, 39–42% of the variation was at the herd level. In the present study we had no information about the farmer; rather, their influence can be considered to be part of the herd effect. Thus our results are in line with the results in Vaarst, who found that the choice of veterinary treatment was influenced by the farmer, as they put different weight on a number of cow characteristics. A recent study has evaluated the extent to which mastitis incidence could be explained by farmers' behaviour and attitude. It was found that self-reported behaviour and attitudes combined explained 29% of the variation in clinical mastitis between herds. Further, the culling strategy, the number of person-years devoted to dairy herd management, and the treatment strategy after the observation of single clots have been found to influence the incidence of clinical mastitis. Differences in the thresholds for treatment and the choice of diagnoses have also been found in interviews with practising veterinarians who treat cattle, indicating that veterinarians are another source of variation.

Our results show that a substantial part of the variation in the odds of veterinary treatment concerns variation between herds. Hence, future studies are also needed to explore effects of the characteristics and attitudes of farmers and veterinarians, as well as the expected economic value of different treatment strategies.

Preventing Bias

In the present study, only one (randomly chosen) diagnostic event per animal was included and this event was only taken from herds with at least four diagnostic events in either cows or young animals. The reason for this is that small groups (few events per herd in our study) have been shown to result in biased estimates in multi-level models. For example, when evaluating bias in clustered data, Clarke found that two-level models with a group average of five produced unbiased estimates both for fixed and random effects. With a group average d" 2, the two-level model produced a downward bias in fixed effects and an upward bias in the random effect.

In the present study, the models were estimated using three different estimation procedures: the LL, the PQL and the MCMC procedures. The estimates from the different procedures were similar for the cow model and for the fixed effects in the young animal model. Multilevel models with binary outcomes have been shown to produce biased estimates and the different estimation procedures used has different advantages, and disadvantages. While it has been found that the MCMC procedure produces less biased results, the LL procedure is preferable during model building since the contribution of the potential explanatory variables can be evaluated using the likelihood ratio test. Further, the PQL procedure is preferable for the evaluation of the model fit. Because the methods available in multilevel modelling, to the authors' knowledge, have different benefits and limitations, several estimation procedures could be used to ensure confidence in the results in multilevel modelling. This is an approach that has been adopted previously.

In studies based on disease events reported by veterinarians or cattle owners, a misclassification (that animals are classified as healthy when there are no records saying they are diseased) bias is likely to be present. In our previous work the proportion of events missing in the farmers' data, but reported to the DDD by veterinarians, was 0.12. We found it likely that those farmers who failed to report events to us that had resulted in veterinary treatment, also, to a greater extent, failed to report events that had not resulted in veterinary treatment. We therefore defined criteria for what we thought was an acceptable loss of data and included the variable 'good record-keeping ability' in the analysis. This variable was significantly associated with the OR for veterinary treatment in the young animal model, but not in the cow model.

As discussed previously, events resulting in veterinary treatment may be easier to recall because the farmers have a copy of the veterinary record. It is therefore more likely that the farmers forgot to report events where the animal did not receive veterinary treatment than events that did result in veterinary treatment. This could have affected our results and more studies are needed to validate our results and assess the extent of this recall bias. The farmers gave, in general, a more detailed description of the disease event if a veterinarian was not contacted, indicating that the farmers were relying on the veterinarians' diagnosis in cases where the veterinarians was contacted. The thorough description of disease events not receiving veterinary treatment enabled the further categorisation of disease events diagnosed as "other disease" by the farmer.

The Risk yo Your Animals

Tuberculosis (TB) is a serious disease of warm-blooded mammals arising from infection by Mycobacterium tuberculosis (MTB) complex. This is part of a group of closely related bacteria that includes:

- Mycobacterium bovis (M. bovis) (responsible for TB in cattle and other mammals)
- M. tuberculosis (the primary agent of TB in humans)
- M. bovis BCG (an attenuated strain of M. bovis used as a vaccine against human TB)

Bovine tuberculosis (bTB) is the disease in cattle that results from infection with *M. bovis* bacteria, and is one of the most complex animal health problems currently facing the farming industry in Great Britain. It is a notifiable disease and suspicion of the disease must be reported to your local Animal Health office.

Bovine TB is a chronic disease and it can take years to develop. *M. bovis* grows very slowly and only replicates every 12-20 hours. The lymph nodes in the animal's head usually show infection first and as the disease progresses lesions will begin to develop on the surface of the lungs and chest cavity. Due to the slow progression of infection, and the Government's compulsory testing and slaughter programme clinical signs of bTB, such as weakness, coughing and loss of weight, are now rarely seen in cattle in GB. Most cattle herds are tested for bTB at least every four years which identifies most infected cattle before clinical signs of disease become apparent. *M. bovis* can also infect and cause disease in badgers, deer, goats and many other mammals, including people. Cases or suspicions of bTB in other species should be notified to your local Animal Health office, although there is no statutory routine testing programme for the disease in other species.

- Information leaflet: Tuberculosis in deer
- Information leaflet: Tuberculosis in mammals

How is Bovine TB Spread?

There is still some uncertainty surrounding bTB and the way it is transmitted though it is spread primarily through the exchange of respiratory secretions between infected and uninfected animals. This transmission usually happens when animals are in close contact with each other. Bacteria released into the air through coughing and sneezing are inhaled by uninfected animals and the disease is able to spread. Cattle to cattle transmission is a serious cause of disease spread which is substantiated by scientific evidence.

The evidence for a link between bTB in badgers and bTB in cattle was reviewed in 1997 by the Independent Scientific Review Group, led by Professor John Krebs.

Information leaflet: The disease may also be spread by contaminated equipment, feedstuffs and slurry.

Recognising the Disease

What to Look for?

Due to the slow progression of infection, and the Government's compulsory testing and slaughter programme clinical signs of bTB, such as weakness, coughing and loss of weight, are now rarely seen in cattle in GB. Most cattle herds are tested for bTB at least every four years which identifies most infected cattle before clinical signs of disease become apparent.

Clinical Signs

During the early stages of infection animals will often show no signs of disease. In the later stages, common signs include progressive emaciation, a low grade fluctuating fever, weakness and inappetence. Some animals will exhibit a moist cough that is worse in the morning, during cold weather or exercise. In the terminal stages, animals may become extremely emaciated and develop acute respiratory distress.

In some animals, the retropharyngeal or other lymph nodes enlarge and may rupture and drain. Greatly enlarged lymph nodes can also obstruct blood vessels, airways, or the digestive tract. If the digestive tract is involved, intermittent diarrhoea and constipation may be seen.

Post-mortem Lesions

Bovine tuberculosis is characterised by the formation of lesions (tubercles) where bacteria have localised. Some tubercles are small enough to be missed by the naked eye, unless the tissue is cut. In cattle, tubercles are found in the lymph nodes, particularly those of the head and thorax. They are also common in the lung, spleen, liver and the surfaces of body cavities. In disseminated cases, multiple small lesions may be found in numerous organs.

Due to the early detection of infection as a result of disease surveillance (testing) infected cattle typically have few, if any, visible lesions at post mortem examination.

Chapter 4

Reducing the Risk of Disease

Good husbandry practice helps to reduce the risk of livestock becoming infected with bovine TB. Best practice would include:

- Provide good ventilation in livestock housing and do not overcrowd your animals
- Keep livestock away from freshly spread slurry and dispose of bedding so that cattle cannot gain access to it
- Work with your vet to formulate a health plan for your herd
- Keep animal identification and movement records accurate and up to date so that cattle movements between herds can be quickly traced.
- Keep your livestock away from neighbouring livestock
- Know where bought-in animals have come from, and their health status
- Keep bought-in and returning stock separate from other livestock
- Keep wildlife out of farm buildings
- Practice good biosecurity.

Biosecurity

Good biosecurity practice helps to minimise the risk of disease occurring or spreading, safeguarding the health and welfare of animals and protecting the viability of businesses.

Good Biosecurity, which literally means 'safe life', provides:

- Peace of mind, healthy stock and a more viable business
- Protects your neighbours and the countryside

- Keeps new disease out
- Reduces the spread of disease
- Keeps more animals healthy
- Cuts costs of disease prevention and treatment
- Improves farm efficiency.

The message is simple:

- disease may not always be apparent, especially in its early stages
- be clean, particularly if handling animals or moving between different premises
- a good biosecurity routine is always essential – not just when there is a exotic disease outbreak, it helps to protect against endemic diseases too.

It is the responsibility of everyone coming into contact with livestock, or the environment in which they live, to play their part in helping to avoid the devastating effects of animal disease.

Good biosecurity should be routinely adopted as part of farm management to help reduce the risk of exotic and the burden of endemic disease. Biosecurity plans should be part of any herd or flock health plan.

Good Biosecurity Practice

- Buying new stock / Returning your stock to the farm
- Hygiene
- Clean food and water
- Separation and isolation
- Slurry Management
- Traceability and identification
- Safe disposal of fallen stock and animal by-products

Buying New Stock and Returning Your Stock to the Farm

Always know the health status of animals you are buying or moving!

- Incoming and returning stock should be kept separate from the rest of the herd/flock. Discuss with your vet and agree a testing programme
- Use separate equipment and staff or handle isolated stock last
- Keep isolation buildings as near as possible to the farm entrance and separate from other livestock buildings by 3 metres

- If using a paddock, keep it separated by at least 3 metres (with double fencing) from other animals on the farm
- Dispose of bedding so other livestock can't have access to it.

Buying in animals carries a risk of introducing disease. There is a hierarchy of risk when sourcing animals; there are no absolute guarantees of freedom from disease but it is possible and worthwhile to determine the degree of risk. The sources of replacement livestock influence the degree of risk.

Animals in the lowest category of risk are:

- Animals from health scheme herds certified free of specific diseases
- Animals from health scheme herds being monitored for specific diseases
- Single source herds of known disease status
- Animals from multiple sources of known disease status
- Single source herds of unknown disease status or in which the disease is known to have occurred.

Those in the highest category of risk are:

- Animals from multiple sources of unknown disease status or in which the disease is known to have occurred.

There are key practices that should be routinely employed to reduce the risk of disease occurring or spreading on the farm.

Purchasers should:

- Be aware of which diseases pose the greatest risks.
- Know what questions to ask of prospective vendors to find out the likelihood of a disease being present (for example, ask for the date and result of the last TB test carried out on the herd in question or encourage the vendors to have a TB test completed prior to purchase).;
- Consult their veterinary surgeon and develop an overall health plan (this is an opportunity to make a long term plan for incorporation in to the farm management system).
- Inspect animals, preferably on the farm of origin, before purchase where possible.
- Keep incoming animals separate from the main herd or flock for an appropriate period (including where animals are acquired from herds certified free of specific diseases). Legislative requirements for separation and isolation exist under the

Animal Disease Control Movement Restrictions. Regular careful inspection of segregated livestock should take place throughout the separation period.

- Preferably purchase directly from the farm of origin and avoid mixing of animals during transportation.
- Be aware that if you buy in disease you may put your neighbours at risk.

Vendors should:

- Develop a health plan with their veterinary surgeon.
- Have information for prospective purchasers on the disease status of the herd supplying animals for sale.
- Provide details of their latest TB and brucella test or have the herd tested prior to sale.

Hygiene

Biosecurity plans should be part of any herd/flock Health Plan. Good hygiene requires constant effort by all who deal with farm livestock. The following guidelines will help to prevent the spread of animal diseases:

Disinfectants

- Only approved disinfectants should be used and dispensed in accordance with specified dilution rates and labelling instructions.

People

- Train staff in the principles of hygiene and disease security.
- Include signs directing visitors to the farmhouse/office and urging visitors not to feed animals or get in close contact.
- Reduce the number of visitors to your farm; consider having a farm post box at the end of your drive.
- Where possible a hard standing area away from livestock should be provided for visitors' vehicles.
- Keep farmyard and surroundings clean and tidy to discourage vermin.
- Provide cleaning and disinfectant materials for all visitors/ workers.
- Consider offering protective clothing/footwear.
- Wellington boots are the recommended form of footwear on farms because they are easy to clean and disinfect. Use a hand

held brush to clean material off the surface and place the boot/sole in the disinfectant solution. Brush away from the face/eyes and avoid contact with the skin. Important: Disinfectant is ineffective if dirt is present, therefore thorough cleaning is necessary before disinfectant is applied.

- Wash your hands with soap and water after handling livestock.
- Avoid wearing dirty clothes and footwear off the farm. This is particularly important when visiting markets, shows, farms and other premises where there are livestock.

Buildings, Equipment and Vehicles

- Vehicles should be kept clean inside and out.
- Clean and disinfect vehicles and trailers (preferably with a power hose).
- Pay attention to areas where dirt may be 'hidden', e.g. wheel arches.
- Clean and disinfect all shared and hired equipment before and after use.
- Animals kept indoors should regularly have fresh clean dry bedding added. Used bedding can cause contamination through urine, faeces, blood, etc. and so should be disposed of away from livestock, humans and watercourses.
- Clean and disinfect buildings and equipment after use by livestock.
- Used equipment, e.g. disposable clothing, veterinary treatments (such as syringes), should be disposed of safely.

Feed and Water

Various diseases can be spread by contaminated feed and water. The risk is reduced by:

- using mains water wherever possible.
- have water bowls or drinkers above the level for faecal contamination.
- avoid contamination of watercourses.
- clean feed and water troughs regularly.
- discourage dogs and cats from walking in feed troughs.
- keep feed in a clean, dry store.
- keep feed stores covered and shut to ensure no access by dogs, cats, vermin and wildlife.

- dispose of old or soiled feed safely.
- Swill feeding is banned.

Separation and Isolation

The following points are a basic guide to basic good separation and isolation practice:

- Keep new animals separate from the rest of your herd/stock until your vet is sure they carry no disease. In many cases 13-20 days depending on animal, will provide sufficient time for disease to become apparent. In other cases, for example brucellosis, incoming breeding heifers may need to be kept apart from the main herd until they have calved normally, up to a year after purchase.
- Minimise nose to nose contact with neighbouring stock.
- Have stock proof boundaries.
- Check boundaries regularly.
- Prevent animals from straying onto roads.
- If common grazings are used try to isolate incoming stock before turning them onto common grazings.
- Keep visitors and their vehicles away from livestock.
- At the first sign of illness isolate sick animals, with the dam if appropriate, and check all the other animals in the herd/flock etc. Thereafter, handle isolated stock last.
- Newly born animals are particularly susceptible to disease so make sure that designated calving and lambing areas are regularly cleansed and disinfected.

Slurry Management

Infections can survive in slurry and manure. To reduce the risk:

- spread on arable land rather than grass for silage making or stock grazing. If this is not possible, allow a 6 week gap between spreading and access by livestock.
- spread slurry using an inverted spreading plate.
- avoid using hired or shared spreaders if possible. However, if sharing is unavoidable, ensure spreaders are thoroughly cleansed before coming on your farm and again before leaving.

Traceability and Identification

There are legal requirements for the registration of all livestock and premises, animal identification, breeding and movement records. Rapid traceability is important for effective disease control.

Keepers of cattle, sheep, pigs, goats, poultry and farmed deer must be registered and their animals identified in accordance with the law, even if only one animal is kept.

Vehicles and people can spread disease. Veterinary investigations into notifiable disease would be greatly assisted by records of visitors and deliveries.

Safe Disposal of Fallen Stock and Animal By-products

Animal by-products (ABPs) are animal carcases, parts of carcases or products of animal origin that are not intended for human consumption. This includes catering waste, used cooking oil, former foodstuffs, butcher and slaughterhouse waste, blood, feathers, wool, hides and skins, fallen stock, pet animals, zoo and circus animals, hunt trophies, manure, ova, embryos and semen.

The law requires that animal by-products shall be identified, collected and transported and disposed of in a safe manner.**Common Diseases of Meat Goats: Individual Animal Diseases (Proceedings)**

Pregnancy toxemia (twin lamb disease, lambing or kidding sickness) is most common does with triplets, and/or are either thin or obese. Much of the abdominal space is occupied with multiple feti in the uterus during late gestation. Fat accumulation in the abdomen in obese animals also occupies space. Because of lack of space for the rumen, these females have difficulty consuming enough feedstuffs to satisfy their requirements. In late gestation, nutritional requirements increase to 150% of maintenance with a single fetus and 200% with twins. Late gestation is usually during the winter months, when less pasture is available, and as a general rule, poorer quality feeds are available. Pregnancy toxemia is also seen following anorexia secondary to other diseases (ex. foot rot, ovine progressive pneumonia, caprine arthritis encephalitis virus), or with stresses such as bad weather, transporting, etc. There is also a genetic predisposition in some individual animals.

Very early signs of pregnancy toxemia are mild depression, anorexia and possibly limb edema. If left untreated, goats become anorexic and depressed, and soon become recumbent. Neurologic signs including blindness, circling, incoordination, star-gazing, and tremors. Constipation and teeth grinding can also occur increased respiratory rate develops if acidosis occurs. If left untreated, does become recumbent.

Azotemia, both from dehydration and secondary renal disease, is a common finding. A urinalysis will be positive for both ketones and protein. Ketoacidosis is common in small ruminants. Hypocalcemia and hypokalemia may be present due to anorexia. They are not always hypoglycemic. Liver enzymes are usually found to be within normal limits, but occasionally may be increased.

Diagnosis is based on clinical signs, the presence of multiple fetuses, and typical clinicopathologic findings. Differential diagnoses include listeriosis, hypocalcemia, polioencephalomalacia, hypomagnesemia, and meningeal worm infestation.

In small ruminants, very early cases (prior to recumbency) may be treated with oral glucose or glucose precursors. 60-100 mls of propylene glycol orally twice a day, or oral corn syrup or glycerol can be tried. Oral high energy calf electrolytes with bicarb can also be given orally. Rumen transfaunation, vitamin B complex (including B_{12}, biotin, and niacin) are also recommended treatments.

Once females show neurologic signs or become recumbent, treatment must be very aggressive. IV glucose, calcium borogluconate, and bicarbonate may be needed.

Glucocorticoids (15–20 mg dexamethasone) can help by causing gluconeogenesis, increasing appetite and inducing abortion. Prostaglandin ($PGF_{2á}$) should also be used (5-10 mg) for induction of parturition in does. Flunixin meglumine (0.5-1 mg/lb) are indicated if endotoxemia is suspected from dead fetuses.

Removal of the fetuses is critical in these severe cases. Assessment of fetal viability with ultrasound helps with the decision to induce parturition or perform a C-section. Since a breeding date is rarely known, age of the fetuses is hard to determine. If the fetuses are alive, induction of parturition can be an option. However, if the fetuses are already dead, or the condition of the doe is severe, an immediate C-section is warranted. Fluid support during and after surgery is critical. Low birth weights of lambs, kids and calves at the beginning of the birthing season can indicate potential risk of pregnancy toxemia.

Prevention of this disease is through proper nutrition. Maintaining animals in proper body condition throughout the year, and making sure energy and protein levels are adequate in late gestation are important. For does in late gestation, hay should have protein content of at least 10%, and 1-2 lbs of concentrate should be fed per head per day. During periods of stress, particularly cold wet weather, concentrate may need to be increased to 2-3 lbs/head/day (divided in two feedings). Parasite control, disease prevention and decreasing stress are important.

Ultrasonography can help determine which females have multiple fetuses, and these animals separated into groups and fed accordingly.

Addition of an ionophore in a feed or mineral mixture will enhance the formation of the glucose precursor propionic acid, and improve effeciency of feed utilisation.

Ebola Vaccine & Western Gorillas

An Ebola epidemic has had a devastating impact on humans, chimps and gorillas in central Africa over the last decade. There are particular fears for western gorillas (*gorilla gorilla*). Although all apes and chimpanzees are threatened, these gorillas have a habitat ranging over a particularly small area, with the majority of the population found in parts of Gabon, Republic of Congo and Cameroon.

Outbreaks of Ebola have hit the large, isolated parks, which usually form the core of the gorilla conservation network. In the mid-90s more than 90% of the ape population in Minkébé Park in northern Gabon died during an Ebola outbreak. There were major deaths due to Ebola in Odzala National Park in northwest Congo from 2003 to 2005. Other smaller parks have been similarly affected, and it is estimated that a third of the world's gorilla population living under protection in national parks have been killed in the last 15

The Ebola Virus

The Ebola virus causes Ebola haemorrhagic fever. Symptoms include vomiting, diarrhoea, general body pain, fever and both internal and external bleeding. The mortality rate is high: around 80% in humans and over 95% in gorillas. The cause of death is usually hypovolemic shock, where shock is induced by a sudden loss of blood-volume, or organ failure. Illness occurs from 2 to 21 days after infection, but in most cases the time between symptom onset and death is 7 to 14 days. The first outbreaks of the virus occured simultaneously in Zaire (now the Democratic Repuplic of Congo) and Sudan in 1976, since then Ebola outbreaks have occurred ebruptly, then spread from person to person. The pattern of outbreaks, both in humans and non-human primates, suggests that the virus is carried by a small, abundant carrier species.

Fruit bats are considered the most likely candidate, but despite efforts of the World Health Organisation in identifying the cause of the disease, this has not yet been shown conclusively. Outbreaks in humans are also triggered by contact with dead primates, and reducing contact between humans and wildlife may reduce the spread of the virus.

The cost of the disease is great. It kills many people in the areas affected, and is destroying primate populations. Many of the regions around the conservation areas rely on eco-tourism for income, and the impact on the gorilla and chimpanzee populations has a direct effect on local livelihoods.

In March 2003 an emergency meeting was held in Brazzaville, Republic of Congo to address the Ebola virus outbreak, both immediately and in the long term. This was the first multi-disciplinary gathering of experts to discuss Ebola in terms of its impact on both people and wildlife. Measures were put in place during this meeting to ease the social and economic traumas of the disease, which threatens both biodiversity and human health.

Present Impact on Western Gorillas

Over the last decade the virus has spread across equatorial Africa at a consistent rate of around 50 km per year, and it seems likely that this will continue. If the gorillas in the parks which are currently unaffected by the virus suffer the same mortality rates as those seen so far, Ebola may kill a further 15% of the world's gorilla population over the next 10 years.

While the Ebola virus alone does not threaten apes and chimpanzees with extinction, this epidemic has reduced the population to a point where it can no longer sustain itself in the face of poaching and other pressures. Gorillas breed slowly, and it is estimated that even with optimistic reductions in disease and hunting, it would take the population 75 years to recover from its present situation. The World Conservation Union (IUCN) raised their status to critically endangered in September 2007. This is the first time that a mammal has become critically endangered as a direct result of disease.

Potential for Vaccination

As most transmission is through small carrier species, which would be difficult to control, it is unlikely that transmission can be completely prevented. However, recent studies show that once the virus has reached a population of gorillas it spreads from one to another, even moving between their social groups. In these situations a vaccination programme could be highly effective, breaking the chain of transmission. There are relatively small numbers of apes remaining in equatorial Africa to be vaccinated, and since the gorilla population is falling so quickly, even saving a few thousand in key areas would make a difference.

Several vaccines have protected laboratory monkeys from Ebola, and they are likely to also be effective in chimpanzees and gorillas. It is hoped that as well as saving western gorillas from extinction, these will also help pave the way for a human Ebola vaccine. The vaccinces are now being developed by a collaborative task force, whose objective is to tackle the ape Ebola crisis. To adapt each of these vaccines for use in apes their efficacy must be researched, and considerable safety testing is needed. Captive and field studies will be used to determine the best delivery methods for the vaccine. The two possible delivery methods available are intramuscular injection by darts, or an oral vaccine delivered by baiting.

Oral vaccines have controlled rabies across Europe, and are presently being developed as part of a successful vaccination programme to save the endangered Ethiopian wolf. Similar baiting programmes could reach a large number of gorillas, but require research to find a successful bait, which is attractive to only the target species. Oral delivery methods involve live vaccines, and so need extensive testing before they can be introduced to wild populations.

There are fewer safety concerns over darted delivery methods, as the vaccines are not live forms of the disease. The darted vaccines will be available quickly as they need less rigorous testing, but far fewer gorillas can be immunised this way. A successful vaccination programme will require development of both delivery methods. These could be used both to control transmission in affected areas, and immediately ahead of the epidemic wave.

It is hoped that the recent findings concerning the spread of Ebola, and the 'critically endangered' status of western gorillas will bring the funds needed to complete the testing of an effective vaccine, and to research suitable delivery methods. The current cost of field and laboratory testing required to launch a successful vaccination programme is currently estimated by the IUCN to be around $2m.

Veterinary Procedures Used to Diagnose Joint Disease in Animals

Information from a number of sources is often necessary to diagnose joint disease in cats and dogs. Some diagnoses can be quite evident, while others may require sophisticated testing and procedures.

History

The veterinarian needs an adequate history of the problem to aid in the diagnosis of arthritis and joint disease. Questions commonly asked include:

- What signs of disease have you seen?
- What limitations is the pet showing, e.g., can it jump into the car or onto a chair, go down steps, put weight on all limbs?

- Did the condition start suddenly or over a period of time?
- How long has it been going on?
- Since you first noticed it, is the problem getting better or worse?
- Is the problem worse or better after exercise?
- Has more than one limb been affected?
- Is the pet showing any other signs of illness besides those related directly to the joint, e.g., is the pet eating and drinking normally?
- Does the animal have a specific use, e.g., field dog, show animal, sled dog?
- Do you know of any injuries the pet has had?
- What previous illnesses or medical conditions has the pet had?
- If the pet is young, are other littermates affected?
- What vaccinations has the pet received, and when?
- Is the pet taking any medications or supplements?
- What does the pet eat?
- Did the parents of the pet have any muscle, nerve, joint, or bone disease?

Physical Exam

The veterinarian will examine the whole animal. Do not be surprised if the veterinarian examines all the rest of the animal before concentrating on the affected area. Mistakes can be made, if we do not look at the entire animal. It is also helpful to start examining the nonpainful areas and then move on to those that show discomfort.

The exam will generally include:

- Weighing the animal
- Taking the temperature
- Listening to the heart and lungs
- Checking eyes, ears, nose, etc.
- Examining the skin for any trauma or puncture wounds
- Examining all the limbs
- Watching the animal move about the exam room or outside on the grass
- Performing special manipulations of various body parts, e.g., neck, limbs

If the pet is in a lot of pain, the veterinarian may suggest lightly anesthetizing the animal so a more complete exam can be made.

Radiographs

In almost all cases of joint disease, radiographs (x-rays) can provide valuable information to confirm a diagnosis, assess the extent of the injury, and determine the proper treatment.

Many times, the changes in the joint may be extremely subtle. Therefore, it may be necessary to also radiograph the opposite 'normal' leg, anesthetize the animal so he is perfectly still when the radiographs are taken, take the radiographs of the joint from different angles, inject dye into the joint and then retake the radiograph and/or submit the radiographs to an expert veterinary radiologist for interpretation. Sometimes, special radiographic equipment which is available only through university veterinary teaching hospitals or referral centres is needed.

In addition to radiographs, other procedures such as ultrasound, CAT scans, and MRI's are sometimes used.

Complete Blood Count (CBC)

A complete blood count (CBC) examines the different types of cells in the blood. This can give the veterinarian important clues on what may be causing a lameness. For instance, with a bacterial infection, certain white blood cells which help fight disease are often increased.

Chemistry Panel

A chemistry panel will check for various substances in the blood such as glucose, calcium, and phosphorous. A chemistry panel also includes tests that check the health of the liver, kidneys, and other organs. Chemistry values would be abnormal in diseases such as diabetes mellitus, hyperparathyroidism, and some infectious diseases.

Serologic Tests

Serologic tests are tests performed on the liquid portion of the blood to check for various immunologic diseases (e.g., rheumatoid arthritis) and infections (e.g., Lyme Disease). These tests are normally not included in the CBC or chemistry panel, and often have to be sent to a special laboratory to be performed.

Synovial Fluid Analysis

In some joint diseases, especially if there is swelling in the joint, a fine needle may be placed into the joint and a small amount of the joint (synovial) fluid is removed with a needle and syringe and then analysed. The analysis will help determine if there is an infection, an immune system abnormality, or if trauma may have occurred. If there is a bacterial infection present, the fluid can be cultured to determine the type of bacteria present. Tests can also be done to determine which antibiotic would best kill that type of bacteria.

Examination of the Joint and Biopsy

In some instances, the above tests do not provide the information we need to make a firm diagnosis. In those instances, a biopsy, or small piece of the tissue lining the joint is removed and examined. A biopsy is especially helpful in determining if cancer or inflammation caused by an abnormal immune system (such as in rheumatoid arthritis) is present. The biopsy may be obtained through arthroscopy, in which an endoscope is inserted into the joint through a tiny incision. Arthroscopic examination and surgery are being used more and more in human and veterinary medicine. Using an arthroscope, the veterinarian can examine the inside of the joint and obtain biopsy material.

In some cases, the joint may actually be surgically opened. This is called an 'arthrotomy,' and is used in small dogs and cats, whose joints are too small to examine arthroscopically. During an arthrotomy the joint can be examined, a biopsy taken, and repairs to the joint can be made, if necessary.

Manual on Livestock Disease Surveillance and Information Systems

The FAO has always been concerned with agricultural development and food security. Recent disease epidemics, in both developing and industrialised countries, have once again focussed attention on livestock disease and their potential to harm development.

In the context of developing countries, disease epidemics do four things:

- They reduce herds and flocks dramatically, which, in the case of pastoral peoples, is a major blow to food security and the ability to survive;
- They cause trading partners to - quite understandably - put trade barriers in place in order to protect their own countries from infection. Where livestock or meat exporting countries are affected by epidemics, their "pariah" status can cost millions of dollars in terms of foreign exchange losses, and drive farmers and the local meat industry to the wall.
- They are a deterrent to sustained livestock production.
- They add significantly to the cost of livestock production through the necessity for the application of costly disease control measures.

Defining the Importance of Diseases

When does a disease become important enough to warrant official intervention? Or to merit international attention? Much attention has been given to highlighting this issue in recent years. The International Office for Epizootics (OIE) has classified animal diseases into two "lists" - List A and List B in order to characterise their level of significance in terms of international trade.

The most important diseases are classified under LIST A. The definition of List A diseases is:

> *"Transmissible diseases which have the potential for very serious and rapid spread, irrespective of national borders, which are of serious socio-economic or public health consequence and which are of major importance in the international trade of animals and animal products".*

List A diseases are:

Foot and mouth disease	***Vesicular stomatitis***
Swine vesicular disease	Rinderpest
Peste des petits ruminants	Contagious bovine pleuropneumonia
Lumpy skin disease	Rift Valley fever
Bluetongue	Sheep pox and goat pox
African horse sickness	African swine fever
Classical swine fever	Highly pathogenic avian influenza

Newcastle Disease

Of lesser importance are the LIST B diseases. Their definition runs as follows:

> *"Transmissible diseases which are considered to be of socio-economic and/or public health importance within countries and which are significant in the international trade of animals and animal products."*

This group includes such diseases as: Rabies, Heartwater, Tuberculosis, New and Old World Screw worm, Brucellosis, and many others.

Recent events such as the BSE epidemic in Europe and the outbreaks of Nipah virus in Malaysia shown that even "unclassified" diseases can have severe economic or trading implications, especially when there is a link to public health.

FAO/EMPRES: A New Emphasis

On taking office in January 1994, the new Director-General of FAO decided that the Organisation should be better focused in championing the goal of enhanced world food security and the fight against transboundary animal diseases and plant pests as outbreaks of such diseases or pests can result in food shortages, destabilise markets and trigger trade measures. A new programme with two sub-components was established: one to combat plant pests and diseases, and one to fight livestock diseases. These programmemes fell under the umbrella of EMPRES - Emergency Prevention Systems for transboundary diseases of animals and diseases and pests of plants.

This put livestock diseases something of a different light: transboundary diseases were now a specific target, and they are defined thus:

> *"Those diseases that are of significant economic, trade and/or food security importance for a considerable number of countries; which can easily spread to other countries and reach epidemic proportions; and where control/management, including exclusion, requires co-operation between several countries".*

EMPRES has classified transboundary animal diseases into three flexible categories. These are:

- Epidemic diseases of *strategic importance*, namely rinderpest, foot-and-mouth disease and contagious bovine pleuropneumonia (CBPP) - these are accorded top priority by EMPRES at the global level. However, regions or countries can have a country-/region-specific set of strategic diseases, as well.
- Diseases requiring *tactical attention* at the international/ regional level, e.g. Rift valley fever, lumpy skin disease, Peste des Petits Ruminants (PPR), Newcastle disease, African swine fever (ASF) and classical swine fever
- *Emerging* or *evolving* diseases, e.g. BSE, porcine reproductive and respiratory syndrome (PRRS)

The diseases in the first two groups - diseases of strategic and tactical significance - have a particular danger in that their occurrence can evolve into epidemics which may threaten populations in a region, and have dire potential consequences in terms of international trade. When is a disease occurrence an epidemic? This is notoriously difficult to define. One definition given is "*The occurrence in a community or region of cases of an illness, specific health-related behaviour, or other health-related events clearly in excess of normal expectancy*". (J.M. Last)

Where a disease is unknown in an area or has been absent for a long time, only one or two cases may qualify as an epidemic and warrant immediate attention. Where a disease has been present at a fairly constant prevalence level for some time, a marked upswing in the number of cases seen may signal a change in status from endemic to epidemic and will require investigation.

Thus the primary role of surveillance is to detect these changes in status early enough to take action. It means having the ability to detect a new incursion, or changes in present status, and presents a challenge to veterinary services in countries around the world. Renewed attention is being given to Transboundary Animal Diseases (TADs), many of which have their greatest impact in those very countries where surveillance (for many reasons) may be weakest.

Early Detection

The key to success in handling animal disease epidemics is early detection. If a disease can be detected very early in the phase of epidemic development, the possibility exists that it can be arrested and eliminated before it actually inflicts damage. Early detection presupposes that there is a surveillance system in place that will bring infection to light when it is first seen. The country's veterinary authorities are then placed in the position of being able to manage the problem before it becomes uncontrollable, thus protecting the local livestock industry and ensuring food security for those closely dependent upon livestock.

That is why this manual is all about surveillance. Early detection enables early warning and an early reaction. Surveillance is the primary key to effective disease management.

The Need for Surveillance

Surveillance has as its main purpose, early detection of disease. The sooner a disease is found before it makes progress along the epidemic curve, the better. The developing world is full of examples of countries with devastated livestock agriculture and severe economic losses incurred as a result of having found out too late. When the perceived threat of livestock epidemics recedes into the background, and there are spending cuts to be made, official Veterinary Services are usually the first to suffer, with a concomitant loss of ability to detect disease.

Thankfully, there are also examples of countries that did detect the very first outbreaks of disease, and were able to mobilise forces to neutralise them before they spread. It is much easier to tackle a disease problem in a small corner of a country where it is only necessary to deal with a small animal population, than to get to grips with a developing epidemic that is spreading on many fronts.

So much for srveillance and early detection. Surveillance has other roles, as well. One of these is monitoring the spread of a disease in order to manage it effectively. Knowing how fast a disease is spreading, in which directions it is going and the size of the populations threatened are all key factors in resource mobilisation. One needs to know how much vaccine to purchase, how many staff to deploy and where they should be deployed, the length of the cold chain that will be involved, and so on. Even when a disease is not present, but is the subject of regular vaccination campaigns (as in buffer zones), good surveillance will give a good idea of where to vaccinate and how many doses of vaccine to take along.

Surveillance plays an important role in the monitoring of progress in control and eradication programmemes. It is important to have in idea of whether the programme is successful (in other words, whether disease incidence is being reduced) in order to assess the efficacy of the control mechanisms being used. In this sense, surveillance becomes even more crucial during the eradication phases of the OIE pathways for various diseases. In these phases, it becomes necessary to prove the absence of a disease rather than to detect its presence: here, carefully planned surveillance actions are of the utmost importance.

What is Surveillance?

The word "surveillance" has been used by epidemiologists for some considerable length of time, often interchangeably with "monitoring", and it is only recently that serious thought has been given to defining the two words.

Surveillance may be thought of as having a broad definition, in the sense of watching a population closely in order to see if a disease makes an incursion. The object of surveillance is early detection of disease. For the purposes of this manual, a definition of surveillance could then be given as:

> *"All regular activities aimed at ascertaining the health status of a given population with the aim of early detection and control of animal diseases of importance to national economies, food security and trade".*

Monitoring, on the other hand is a more specific activity/ies that will follow as part of an early reaction should surveillance activities indicate introduction of disease. It will focus more specifically on the identified disease in order to ascertain changes in prevalence level, rate and direction of spread. Monitoring can thus be defined as:

> *"All activities aimed at detecting changes in the epidemiological parameters of a specified disease".*

It should be pointed out that many of the techniques used to implement monitoring can be used in surveillance, and vice versa - and in fact, in practice, the distinction between the two often becomes blurred. As this is a practical book, the blurring will be noticeable in the text that follows. Readers will find much that is value in these pages, no matter whether they are anxiously waiting for a disease that they hope will never appear, or nervously following the progress of a disease they wish hadn't broken out. The distinction is more in the objectives than in the techniques applied.

Surveillance efforts, although as all-encompassing as possible are by their nature, are often not planned to be aiming at a particular confidence level in their execution, whereas monitoring is usually mathematically planned and aims to follow disease dynamics with a certain measure of precision. Readers should be aware that doors are open for bias and error in both monitoring and surveillance, and should consult reputable textbooks on epidemiology to ascertain sources of error and bias, and how these are best counteracted.

Mention will be made of many of the pitfalls of the different activities in the text. As mentioned earlier, this is a practical "how to" manual on information gathering and management, and the reader will want to make use of other texts for a fuller background on many of the concepts introduced here.

Surveillance on the Ground—Surveillance Planning

Although perhaps not always mathematically precise, disease surveillance is not a haphazard action, but a meticulously planned and managed activity. It involves the deployment of personnel who will be moving in the field in a carefully programmed manner, using various methodologies to detect signs of livestock disease. This manual assumes that since the control of disease epidemics is in the greater public interest, it will be official veterinary staff who are involved in doing the work of surveillance for those diseases. Obviously, not all governments will have large armies of veterinary personnel; staffing levels will vary according to budgetary constraints and the importance of the livestock industry; this manual endeavours to deal with such differences.

Putting a Surveillance System in Place

National governments must realise that animal disease surveillance is a key function of their national veterinary services. Once that is an accepted principle, there is a need to agree on the need for a property structured and administered system. Having a properly implemented and utilised system will require co-operation from a number of stakeholders - the official veterinary service (management and field staff), other extension staff, private veterinarians, farmers and other organisations that might be operating on the ground, for example NGOs. Assuming at this point that it is the official veterinary service that is initiating the drive, it will have to make contact with these stakeholders, explain the intention, and enlist their support. It will also require transparency, as those who are involved in data supply and collection will want to see that the information they supply is actually put to use.

In developing countries, properly supervised sub-professional groups (veterinary assistants, auxiliaries and community animal health workers and the like) are often important elements in surveillance systems, and must be singled out for special training.

The next item on the agenda would be to agree on the objectives of such a system. A specimen set of objectives might be:

- the early detection of livestock diseases of economic/food security/public health importance
- enabling early reaction to such diseases
- correct identification of resource needs in the field so that existing resources can be correctly deployed in disease management.
- provision of strategic decision-making support
- measurement of surveillance system performance.

What would the priorities of such a system be? The first would be to identify the most important *diseases* in the country. Field staff (and farmers, for that matter) would need to be familiarised with these diseases, and at least know the basics about recognising them in order to be able to report their presence. There is obviously nothing wrong in reporting any disease through the system, no matter how insignificant it may seem; the problem comes in allocating resources for staff training - this is where decisions will have to be made about priority diseases. Time and training materials will have to be devoted to priority diseases.

Having identified priority diseases, the next step is to identify *priority areas*. It will probably also not be possible to direct resources equally throughout the country, and the areas where the identified diseases pose the greatest problem will have to be earmarked for the first resources in terms of training, staff, intensity of surveillance, etc.

It is likely that priorities will differ in different parts of a country, and that must be taken into account. While CBPP may be of importance in one province, trypanosomosis may be of greater importance in another. Planning of training and awareness creation will have to take such differences into account.

Beyond a country's borders, it may be that other diseases enjoy priority. In each region of the world, a different set of circumstances, with its own set of livestock diseases, exists. Being a responsible global citizen will require a country's veterinary service to be aware of regional priorities in terms if disease surveillance, and internal surveillance systems will also need to take account of these. Regional economic and political groupings often play a leading role in drawing up regional priority lists of animal diseases.

A final criterion in developing a priority list for surveillance is that of international disease reporting. Every country that is an OIE member will need to decide what OIE List A and B diseases are of local importance, how to detect these diseases, and how to report on them. Transboundary Diseases as defined by the FAO will also play a role in shaping surveillance systems. Defining priorities is therefore not purely a matter of local interest, but also a matter of sensitivity to international concerns. The greater the attention given to diseases that are transboundary in nature, the greater the opening for participation in international trade and economic advancement.

Targets must also be defined, as well as responsibilities. In the various priority areas, decisions will need to be made concerning staff development levels, and time-frames set. What will the frequency of surveillance inspections be in area? When will they start? If there is to be sero-surveillance, when will the first survey take place? Will there be regular serosurveys? Who will be responsible for each activity? Will computers be used? Do questionnaires need to be drawn up?

A proper management plan, defining priorities, targets, responsibilities, resource allocation and responsibilities must be drawn up and adhered to. The various elements that should be included in such a plan will be made clearer in the pages of this manual.

In drawing up a management plan, the normal information flow in a surveillance system will need to be taken into account, and every step in the flow properly monitored and controlled.

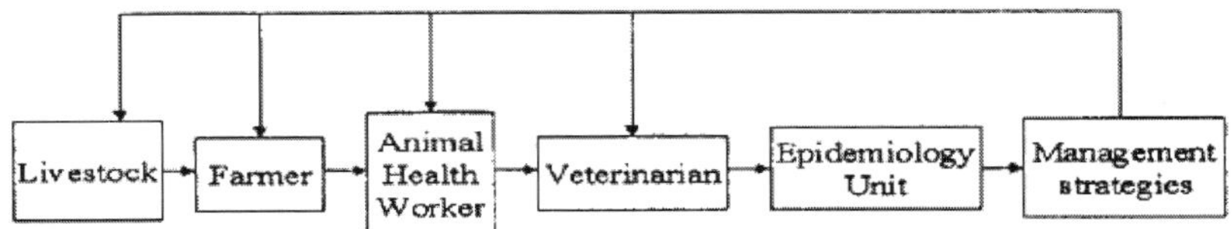

Figure: *Information flow in Surveillance and Livestock Disease Management*

Surveillance for What?

If information is to be collected, it will have to be managed, and - most important - *used*. Data collection with no clear purpose is a waste of time and resources. One thing that data managers will have to define very clearly, is *what incident will trigger what action, and at what level.* The data that should be collected is that data which will lead to action.

For example, the first-time suspicion of a Transboundary Disease in an area will require immediate action by the local veterinarian - perhaps in terms of quarantine or movement control, and certainly in terms of follow-up. At management level, it may require further decisions in terms of publicity and redeployment of resources.

Exactly what action is taken by who, and under what circumstances, must be spelled out as a part of the surveillance system. It will obviously differ from country to country, and even from region to region within countries, and this type of planning rests firmly in the domestic arena. But it is planning that *must* be done.

Surveillance When and How?

It was mentioned earlier that surveillance is a planned activity, involving a carefully laidout routine. The type and level of activity, the routine (frequency of activity in any area) are all determined by the country's veterinary management, in agreement with the various stakeholders, according to the criteria listed above.

Suffice it to say that whether government staff are combing an area in large numbers looking for a disease themselves, or whether a small team is conducting retrospective farmer interviews, the activity must be a planned activity that is worked into the annual activity plan.

A little extra time and resources spent on surveillance at the beginning will save a lot of time and money later on should a disease break out. Surveillance planning will include such things as deciding on how many staff members will be devoted to surveillance, the frequency of visits, distances to be travelled and transport requirements. Planning must be such that activities are evenly spread, not that one day is overburdened with work (creating the necessity for a "rush job") while another day's activities are very light. This kind of information is necessary not only for execution of the activity itself, but also for budgetary planning.

Having a fixed work plan means that the staff member knows exactly where he needs to be and when he needs to be there; and should supervisory staff need to monitor what he is doing, they know where to find him. Should the staff member need to be contacted in the field in an emergency, his whereabouts will be known. If the person is using a vehicle, the distance estimates in the work plan will also serve as a guide to indicate how closely the plan was adhered to.

Visits to the field must be planned in such a way that a single round trip will cover the biggest area possible. Surveillance teams (or individuals) should have areas demarcated for them, and should not overlap. Wherever possible, the same staff should be visiting the same areas every year - this enables the livestock owners to get to know the staff and build up trust in them, and enables the staff to get to know the area and the movements and habits of the people more intimately. Regular surveillance by the same staff member (preferably of the same ethnic group as the farmers being visited) builds mutual confidence and means that farmers will report diseases more readily.

Surveillance does not simply mean that a staff member "does the rounds" inspecting livestock or taking blood samples: it requires that the surveillance officer be involved in a constant process of informing and educating his farmers so that they will immediately recognise telltale disease signs and report their presence without delay. Extension activities could be incorporated into the surveillance plan, and certainly the official should be carrying a supply of information leaflets and/or posters giving details of diseases of local importance.

Surveillance: Local Needs vs National Needs

The battle cry is usually to "keep surveillance focussed". Having a narrow focus for disease control activities will, however, often backfire. Diseases of importance to the national veterinary service are often not of importance at the grassroots level. Low profile rinderpest will almost certainly be of lesser importance to a rural community than concurrent endemic CBPP; ongoing surveillance for FMD in an area where the disease is normally absent (but perhaps occurs only every five or six years) will not gain a community's sympathy. Forcing an approach onto a community won't work, and will only cause resentment. Local communities need to feel that their needs are being catered for, and in order to gain co-operation, compromises will need to be made.

Shaping such compromises means listening to farmers, and taking other disease needs into account. It will certainly mean carrying out surveillance for a package of diseases rather than a single one. The logic for this concept is both economic and practical. The system put into place for surveillance of the critical target disease can easily capture data on other diseases in the same area. The sustainability of the system will be better secured as it will be viewed more positively at both the local and the national level. Furthermore, it should be remembered that even during a disease eradication process, there is a requirement for the surveillance system to demonstrate its capacity for those that are relevant for differential diagnosis.

Nevertheless within a broad-based national surveillance system, there will be a need from time to time for specific targeted surveillance programmes aimed at either specific and special awareness/early warning or in aid of special eradication programmes.

While targeting a broad range of diseases makes sense from the managemental and economic points of view, there is a lot to be said for surveillance "riding on the back" of a particular disease. Very often, it is a single disease (such as rinderpest or Foot-and-mouth disease) which causes a national scare and provides the impetus for putting a national surveillance network in place. Once resources are in place, moving from an emergency response to a proper surveillance system is only a very small step. Moving to wide-spectrum surveillance must ensure that impact scenarios can be constructed. Surveillance also means having the information necessary to predict what the consequences of other disease epidemics might be. Having secured the resources necessary to put a surveillance network in place, the same network must ensure its survival by justifying its economic importance on the national scene.

The same kind of reasoning applies to extension work. If surveillance personnel are equipped for extension (which they should be), then their extension materials should take account of local needs. Training must be provided on the recognition of a number of diseases, so that the network is "sensitive" to a varied repertoire of needs.

This is especially so where a network relies heavily on the reporting by farmers of disease outbreaks. If farmers feel confident about reporting diseases with which they are familiar, they will also feel easier about reporting occurrences that are foreign. The benefits work two ways: farmers feel that something is being done to assist them with their own identified needs, while the surveillance system will have a greater chance of detecting "exotic" diseases.

The Flexibility to Cope with Ad-hoc Needs

Much has been said above about having a planned surveillance schedule that runs for year after year. A routine programme is important, but it must not exclude *ad-hoc* needs. It should be possible to re-deploy staff to take care of new needs - for example, a need for heightened surveillance in a particular part of the country for a short while to guard against a threatening disease incursion. Reviewing surveillance programmes and their efficacy-and, if necessary, changing them to suit population shifts or disease patterns, is an important part of surveillance management.

Visual Surveillance

Having planned a surveillance programme, what types of surveillance could be undertaken? Probably the most popular - and one of the easiest and cheapest - is visual surveillance. Two kinds of visual surveillance can be distinguished: *direct* and *indirect*. For both types, the veterinary official visits all (or as closely as possible to all) farmers in the area allocated to him/her. This is less of a random sampling exercise and more of a census. It is carried out on a regular basis, any number of times per year, depending on staffing, budgetary constraints and the prevailing disease situation.

In *direct* visual surveillance, the observer physically inspects all animals and records what he sees in terms of disease. Probably the best way of doing this with cattle is to first observe them from a fairly close distance and run them slowly through a race/crush if one is available. Individual animals can be selected for closer examination should they appear unhealthy. Small ruminants are probably best looked at from as close a distance as possible in a pen. Again, those appearing sick can be singled out for closer examination. The official's findings are recorded on a pre-printed data questionnaire. It may be possible for the officer to take samples for laboratory confirmation, or even to perform a post-mortem, depending on his/her mobility and other logistics. This approach assumes that the veterinary official involved - who is usually a lay person - has had fairly comprehensive training in recognising diseases of local importance, as well as diseases of national or international importance which might not occur locally on a regular basis.

With indirect surveillance, it is assumed that the livestock inspection frequency is very low, and therefore, while the official visits all possible livestock owners, he/she relies on their recall to gain details of diseases that affected their flocks and/or herds. While some may scoff at this kind of surveillance, it is important to note that most herders have an accurate knowledge of local diseases, and are usually fairly close to target when it comes to making diagnoses. If they aren't always all that accurate clinically (usually because there may be different diseases in an area with similar clinical pictures) there will be certain syndromes which usually have fairly descriptive names in the local vernacular. These names at least - and the approximate numbers of animals which were affected - can be recorded for further investigation if necessary. Usually, by closely questioning farmers about the clinical signs seen, it is also possible to arrive at a tentative diagnosis. Again, it is necessary for the veterinary staff to have a good idea of clinical signs and post mortem lesions to be able get all possible information out of the farmer.

It is important when questioning a farmer not to ask leading questions, or to use a questionnaire with "pre-cooked" clinical signs listed on it (if at all possible). Herders might be led to believe that this is a quiz where "yes" is the right answer, and may well report disease signs that were never seen. In practice, of course, the two types of visual surveillance are never really separated. Usually, the veterinary staff member will inspect all the animals present, record what he sees, and then ask the farmer about all diseases that were seen since the previous inspection. All observations - both direct and indirect, are then recorded.

Sero-surveys

Sero-surveillance, while it may give a more "objective" view of the disease situation in an area (it measures antibodies rather than a lay person's conception of a disease), has its limitations. Among them are:

- Costs - of blood and serum tubes, needles, transport of samples to laboratories and the cost of testing.
- Lay staff will have to be trained in sample-taking.
- Suitable facilities (good crushes with working head-clamps) are not always available in rural areas.
- In some cultures, there may be strong objections to the taking of blood from live animals.
- Animals not always be handleable when it comes to taking samples.
- Because of the costs involved, the survey is more likely to be carried out on a small subset (a random sample) of population than on a larger cross-section of animals, and the question of sampling error then comes into play.
- The specificity (the probability that a non-diseases subject will be classified as diseased) and the sensitivity (likelihood that an

exposed case will be classified as exposed) vary from laboratory test to laboratory test for various diseases, and are never 100%. This also introduces inaccuracies into the surveillance effort.

Nevertheless, sero-surveillance is still more "objective" than visual surveillance, and for many diseases it can be used to get an idea of current prevalence and geographic distribution. The costs and other problems involved should not be a discouragement to use sero-surveillance. A well-planned random survey will carry benefits that should far outweigh its costs. In the case of rinderpest, sero-surveillance can be used for antigen detection (for investigating wild virus activity or proving absence of infection), while seromonitoring is used for post-vaccination antibody detection.

Planning a Serological Survey

First, the objective and the tests to be used must be clearly stated. If surveillance is for Rinderpest, what is being sought? Evidence of exposure to disease, or evidence of vaccination? Where? In what species? Is this part of an eradication process? If so, is it only going to be a zonal process or a national one? Once the objectives have been decided and the area/s to be surveyed have been delineated, other decisions have to be made. What will the sampling units be? The primary sampling units will, in most cases, be herds. Herds, in developing countries, need definition, and these definitions will vary by geography, ethnic group and farming system. It might be a group of animals cared for by an individual, or it might be a group of animals owned by a collection of individuals. The secondary sampling units will then be individual animals. If the individual is important (eg. when trying to link clinical signs to antigen presence or an antibody titre), then identification of the individual becomes important - if not, one needs not be too concerned with secondary sampling units.

Once the type of sampling unit is known, and the survey area is defined, the next step is to draw up a sampling frame from which the psu's will be chosen. The sampling frame would be, for example, a list of all villages in the survey area, together with the livestock populations of each village. The primary sampling units are then chosen by random sampling. Procedure for random sampling will not be discussed here (but a short summary is given in an appendix), as they are described in epidemiology textbooks, as well as FAO and OIE publications (eg. "Recommended surveillance procedures for disease and serological surveillance as part of the Global Rinderpest Eradication Programme (GREP)" IAEA/FAO, 1994). What is important is to determine the confidence limits (usually 95%) and the prevalence level which is desirable to detect beforehand, as this will, in turn, determine the size of the sample to be taken. How to calculate sample sizes is covered in many authoritative texts; it will not be given here. Nomadic herds present a special problem. Very often, instead of doing a random village selection, one can do random selection of grid blocks from a map, enter the areas selected, and then sample a percentage of the animals found inside the block. This technique has been well described elsewhere.

Once the area to be covered, the size of the samples and the whereabouts of the herds to be sampled have all been worked out, the next (and more down-to-earth) stage of planning is reached. This will involve the following:

- Working out the needs in terms of how many teams will be sent into the field, how many vehicles are to be used, a sampling programme, and how many animals each team will have to bleed, how many members are required for each team. Sample tubes and accessories will have to be included as well as marker pens and ear tags (if identification of individual animals is important). Bear in mind extras such cool boxes and aids to animal handling. This will enable a budget to be drawn up for the entire exercise.
- Then comes the purchase of vacuum sampling tubes, needles and needle-holders, and any extras in terms of ropes, nose-tongs etc. Remember when purchasing vacuum tubes that provision will have to made for extra tubes in case of breakages and defective tubes. It is important to bear inmind that serum sampling requires an extra set of tubes for decanting serum from blood (this is often done in the field), and that the total number of tubes required is thus double the number of samples to be taken.
- Tubes must then be marked. If individual animals are important (e.g. in Rinderpest follow-up) a marking code will have to be developed for the village/settlement and the individual. If the individual animal is not important (e.g. in a CBPP sampling exercise where serology is of value in identifying infected herds only), then the village code is of importance on the tubes. This code will have to be used on the blood tube, the serum tube and the sample form. All of this is best done in advance.
- Each team will have to be fully briefed on the programme to be followed, and issued with all items of equipment needed. Small things like pens, sample record forms/questionnaires, clipboards, extra marker pens for sample tubes, etc should not

be forgotten. The individual villages or farms must be clearly identified to the teams, as well as the numbers of animals to be bled at each place. Arrangements must be made for transport and cooling of sample tubes - cool boxes should be provided for.

- If teams are going to be in the field for a number of days, they will need to camp out. Blood samples can be allowed to settle overnight, and the serum decanted early the next morning - alternatively small hand centrifuges (fairly impractical for large numbers of samples) can be used to separate the serum. When sampling teams are spending time in the field, other needs, such as food, camping equipment, extra petrol and essential motor spares must be remembered.

Something often forgotten by field staff planners of serosurveys is to make all necessary arrangements with the laboratory analysing the samples. The laboratory should have the technical ability to perform the tests required, and should be informed of the arrival of the samples well in advance in case extra resources need to be reserved for the job.

"Passive" Surveillance

Most ordinary surveillance routinely carried out falls into the category of passive surveillance. In this case, there are routine programmes that run - usually partly directly visual, or indirect, relying on farmer interviews and notification - basically to survey the landscape for livestock diseases and to detect and changes in status.

This is probably the most important, and is a key element in early warning. The word "passive" should be seen as a characterisation of technique and not a sign of lowered importance of the work done.

"Active" Surveillance

Much has been made of the concepts of "passive" and "active" surveillance. Passive surveillance is usually thought of as regular - and perhaps infrequent - visits to an area by veterinary staff to assess the local animal situation and determine livestock populations. It would include voluntary disease reporting by farmers, traders and perhaps other individuals such as private veterinarians.

"Active" surveillance entails frequent and intensive efforts to establish the presence of disease in an area. Examples:

- A disease threatens from just across a country's border. The threatened country will mount frequent livestock inspections, perhaps coupled with specific clinical observations ("mouthing" for FMD, close examination of mucosae for lesions and discharges for rinderpest). As an adjunct to this, serum samples might also be taken frequently.

- Frequent patrolling of borders or *cordons sanitaire* to detect and follow up illegal livestock movements.
- When a disease is suspected in an area, active surveillance techniques are often used to confirm its presence - or, hopefully, its absence.
- Active surveillance is of great importance in supporting official declarations to eradicate disease. Once the declaration has been made, it is up to country concerned to meet all requirements for proving freedom from disease, and finally, freedom from infection (eg. rinderpest, CBPP).
- Countries may be required, as part of livestock trade protocols, to prove absence of a certain disease (eg. Foot-and-mouth disease, Tuberculosis, BSE).

Any activity which is frequent, intensive and aims at establishing the presence or absence of a specific disease, could be described as "active" surveillance.

Once the presence of a disease is confirmed, and similar techniques are then used to follow trends in its development, this would (at least in terms of current terminology) be called "monitoring".

There is no doubt that active surveillance activities can be expensive and time-consuming. There are benefits, however, that in the long run will outweigh the costs. In the first instance, beginning active surveillance (at least for diseases such as rinderpest and CBPP) means that vaccination has ceased, and huge amounts spent on blanket vaccination campaigns will be saved. Secondly, there are trade benefits to be gained - eventual proof of disease absence will allow the opening-up of hitherto untapped markets.

Abattoirs and Slaughter Slabs

These are a valuable source of data, particularly when it comes to diseases which present laboratory diagnostic difficulties, such as CBPP. It has rightly been said that abattoirs are "the post mortem halls of the nation" and are a goldmine of information. Export abattoirs are always closely monitored by veterinary officials, and instituting a reporting system to extract information from them would present no problem. Smaller abattoirs and slaughter-slabs present something of a problem in that there may not be enough official staff to keep them under surveillance. Possible solutions to this would include the following:

- Entering into agreement with those running smaller facilities - or even butchers at slaughter slabs - to alert the authorities in the event of a possible transboundary disease being detected on slaughter. This will require some investment in terms of basic meat inspection training and the recognition of characteristic lesions, but it may be well worth the outlay.

- Random inspection of slaughter slabs or small abattoirs by officials who might spend a day or half a day keeping watch on the proceedings at a particular place, and then move to other duties - or to the next slaughter slab.
- Intensive inspection activities could be reserved only for those facilities perceived as being in "high risk" areas.

The kinds of information required here would be such things as species, origin of animals slaughtered, lesions seen, condition suspected. During the eradication phases of a disease pathway, surveillance would have to be organised such that:

- it was random
- it covered the required number of animals at a 95% confidence level to give reasonable assurance of the absence of disease.

Rapid Appraisals

Rapid Appraisals go by many names (Rapid Rural Appraisal, Participatory Rural Appraisal, Participatory Epidemiology, Sondeo Method, etc). The basic idea of the Rapid Appraisal is collect information in data-sparse areas, places that are marginalised, remote, inhabited by nomads, inhospitable, infrastructure-poor - whatever adjectives one might choose for neglected areas, whatever the reason for the neglect might be.

Rapid Appraisals provide a means for data collection which would otherwise not be there: the data may be limited in scope, inaccurate or biased - but it *is* data, in certain vast areas of the developing world where information is virtually non-existent, any data are better than none. Such data may not be statistically valid, but will at least provide an indication of what is happening "out there" - and where it is happening. It is based on rural livestock owners' impressions, but their impressions are often very accurate.

Many possible "tools" are available for appraisal work, but the end result is usually an unstructured/semi-structured interview during which the interviewer tries to capture some form of data, at least in a semi-structured form. The essential ingredients are (a) to know exactly what information is required, (b) how to capture it fairly informally, and (c) how to structure it so that it can finally be computerised (if necessary) and analysed.

In summary, Rapid Appraisals make use of the following tools to gather information:

- Background studies of historical, demographic, epidemiologic and geographic information to become acquainted with the area under study (including previous consultants' reports, scientific papers, population census data, official reports and maps).
- Interviews with "key informants" - community leaders, church leaders, government officials, health workers, veterinarians, NGO workers and others in contact with the local people who can provide information on what is happening in an area.
- Group interviews with farmers (need careful organising and chairmanship to ensure fair representation of poor and elite, and to ensure that individuals do not dominate the group).
- Gender analysis - using men to talk to men - to find out what they do, when they do it, and what they know - particularly with reference to livestock diseases. Similarly, women must be used to interview women. This type of technique may not always be acceptable in some communities, but when it is, it should be exploited to the full. Women usually work more with poultry and small livestock, and their knowledge of diseases in this field is extensive and invaluable.
- Direct observation - often, spending a day watching how people in rural areas handle their farming system provides essential information that will give a better understanding of livestock management, and how to handle disease problems. Linkages within the farming system (crop-livestock) may also be better understood. Walking around a farming area will provide valuable impressions of livestock, husbandry methods and the status of the ecology.
- A very good means of carrying out appraisals is the use of community animal health workers or "veterinary scouts". Such people provide a basic service to the community in their private capacity and are remunerated by the community for services rendered (usually medical treatments, or simple procedures such as castrations). Because of their more specialised knowledge, they are excellent sources of information for appraisals. Of course, such people need training and initial drug and equipment supplies at the outset, and this will cost money. Such schemes also do not always work, but in communities where they are self-sustaining, costs are limited to an initial outlay only.

When using appraisal techniques to gather disease data, one must bear in mind the various sources of error and bias:

- Farmer recall. The longer the period between visits to an area, the longer is the recall period required from farmers. This may lead to all sorts of errors in observation, and a group interview - where farmers get the opportunity to validate each others' observations - may be helpful in this regard. Interviews with other informants may also provide a means of cross-checking information, but care must be taken to ensure that informants are really in contact with what is happening at grassroots level, and that their inputs are credible.
- Interviewer bias. Interviewers may have their friends among the more influential farmers and tend towards interviewing them at the expense of poorer farmers. In addition, interviewers might have their own ideas about what is important and may tend to miss the "smaller" details that often make the difference.
- Travel limitations. Restricted ability to travel due to poor roads and inhospitable countryside may miss the more remote areas where any number of disease problems may be brewing. This may also be a stumbling-block to follow up: even if a possible transboundary disease is identified in an isolated area, there may be a reluctance to follow it up and confirm the rumour due to the distance and difficulty of access.
- Community desires for service provision. Farmers may be tempted to exaggerate their disease problems in the hope that this will attract more services from the government.
- Information may "get lost" between interviewer and interviewee if interpreters have to be used.

Despite the fact that Rapid Appraisals are fraught with problems that amount to a statistician's nightmare, they are essential in areas where staff are thin on the ground and frequent and intensive surveillance is not possible. They provide useful information on the *disease situation* in the area, the *farming systems*, the *distribution of people and livestock*, and on *trading routes* - where they run, how and when they are used. Trading routes are notoriously difficult to document, and Rapid Appraisal tools provide a means of recording these routes (which need updating from time to time).

Another important use of Appraisals relates to the provision of baseline *economic data* for the calculation of losses due to disease impact. A number of appraisals, using interviews combined with direct observation in an area, will give an idea of livestock production. Cash values can later be imputed to these estimated production parameters in order to calculate the economic cost of a disease epidemic should it strike the area.

The Role of Laboratories in Surveillance Programmes

A final word here about the role of laboratories in veterinary surveillance. Diagnoses (or tentative diagnoses) can be made in many ways:

- Visual (by non-professional field staff)
- Clinical observation (by veterinarians)
- Post-Mortem evaluations (usually by veterinarians)
- Simple laboratory examinations (blood smears, faecal examinations, impression smears)
- Laboratory methods (serology, tissue culture, bacteriology, histopathology, etc)

Often, laboratories are thought of as having relevance only in serological surveys. However, it is imperative that laboratory backup be obtained for as many diagnoses as possible. It is essential in cases where an epidemic disease is suspected for the first time in an area that samples be taken for confirmation of the diagnosis. Where the diagnosis is uncertain, repeated follow-ups, with laboratory sampling, must be made in an effort to either confirm or exclude the disease.

Where it is known that a certain disease has become endemic, confirmation of each individual focus becomes unnecessary, but 10-20% of cases must always be confirmed to ensure that the epidemiological picture has not changed and that another disease, different in epidemiology but similar in appearance, has entered.

For these reasons, the presence of a strong laboratory diagnostic service, subservient to the official veterinary service, and answerable to it, is absolutely essential. Laboratory services are an essential backup to what is being done by field service staff. It is also necessary - and this is often forgotten - that field staff (including veterinarians) be regularly briefed on the kind of samples needed for the various diseases threatening in an area, and that they are also familiar with the requirements for preserving, packaging and transporting such samples.

Maintaining Laboratory Norms

Laboratory testing needs standardisation so that, for example, the results of serosurveys analysed by different labs are comparable. This means that laboratories need to belong to networks where the same reagents and methods are used in the same test; where experience and expertise is shared; and where use is made of reference laboratories.

A chain of OIE and FAO reference laboratories has been established for this purpose. The FAO/IAEA Joint Division has also been established to assist with standardisation of tests, and for quality assurance. It is imperative that national veterinary laboratories make use of these services.

Surveillance in Resource-poor Countries

Surveillance often presents itself as a thorny issue in developing countries because it is seen as a costly operation necessitating an enormous army of surveillance personnel on the ground. This need not be so, and judicious deployment of resources can often achieve what is needed without great expense.

Critical Point Identification

The first step in setting up a "low cost" surveillance system involves identifying *critical points* or *critical surveillance areas.* These would include:

- areas under direct threat of disease (perhaps due to the presence of a nearby focus)
- border crossings
- watering points or slaughter slabs near migration routes
- auction pens and other major livestock assembly points
- abattoir lairages.

Resource Deployment

The bulk of veterinary resources should then be deployed at these critical points, with high frequency surveillance designed to move staff amongst such points with relative rapidity for whatever type of surveillance is deemed appropriate - visual (for detecting clinical/ pathological signs), detection of antibody, and detection of the causative agent. Such surveillance is fairly structured but not sufficiently randomised for movement along an OIE pathway. It would, however, qualify a country for entrance to a pathway if it gained sufficient evidence of clinical disease absence - the point is that the work would have to be restructured along more "scientific" lines in order to move further along an OIE pathway.

As an aside, it must be mentioned that strategic resource deployment may clash with equity goals - politicians may want to see a more "even" distribution of resources across the country. Careful explanation will have to be given for unequal resource deployment, with the assurance that once the disease problem is cleared up, personnel and equipment will once again be redistributed.

Surveillance Frequency

The frequency of surveillance at these critical points is a matter of common sense and would have to be determined by the perceived risk of each point, with the higher risk points receiving the most frequent attention. Frequency of surveillance will, on the one hand, be determined by the frequency of population turnover (eg. along trade routes) and by the incubation period of the main disease feared at the time. In a relatively tatic livestock population at high risk to foot-and-mouth disease, for example, it would not make much sense for surveillance to be more frequent than fortnightly. On the other hand, financial constraints will also be a major determinant of frequency. Surveillance needs to be an intelligent trade-off between field realities and budgetary limitations.

Non-critical Areas

All other parts of the country would be deemed to be *non-critical areas* where surveillance could consist of relatively infrequent visits by field personnel (perhaps once or twice per year) or annual Rapid Appraisals relying heavily on group interviews.

Useful information can also be gathered from other existing networks - for example NGO workers, crop extension officers who may happen to be in the area, consultants, etc.

An essential item in any surveillance system is *farmer awareness*. Training local livestock owners in disease recognition and encouraging them to report the presence of any suspicious clinical signs is a very cost-effective means of improving the quality of disease surveillance, both in critical and non-critical areas.

There may even be a possibility of a small incentive to be provided for evidence leading to the discovery of a disease - eg. a fee to be paid should a farmer submit part of a diseased lung for CBPP examination.

Data from private veterinarians is an important item not to be forgotten. Capture of this may be via questionnaires sent to them regularly; a legal requirement for them to report certain diseases to the authorities; and by making the use of an official government questionnaire obligatory when sending samples to the laboratory (in this way, data from laboratory submissions will enter the system automatically). Ultimately, the exact type of surveillance adopted by a country is its own decision, based on disease risk and available resources. What is important is the issue of transparency. It is incumbent on each country to make the precise mechanics of its surveillance system known to neighbours and trading partners.

This includes the identification of critical areas and non-critical areas, and the types of surveillance operational in each. Such transparency builds confidence, facilitates mutual risk analysis, and in the long run, will promote investment and trade.

Information Systems

Much will be said about information systems and information management in this manual - for a very simple reason: once information has been gathered, something has to be done with it. Three things must happen to information: firstly, it must be *managed*, controlled and quality-checked; secondly, it mus be *analysed* in order to become more understandable, and thirdly, it must be *acted upon*. These three points will be emphasised repeatedly throughout this manual.

For information to be an analysable and eventually useful for decision-making, it needs careful management and quality control.

What is an Information System?

An Information System is the collection of data, people, procedures, hardware, software, files, and information required to accomplish an organised set of functions.

Tom Adamson

Basically, the constituents of an information system are:

People

- those who gather data
- those responsible for data input
- those responsible for data analysis.

A storage/retrieval/analysis system - which in this day and age would usually consist of:

- computer software
- appropriate computer hardware.

A feedback delivery system:

- mechanisms to ensure that processed data is fed back to data gatherers.

An iformation system is nothing more than a large communication cycle, involving information transmission and reception. In any communication cycle, one person (the communicator) transmits information via a medium (the spoken word, the written word) to a recipient.

In order to ensure that the concept transferred to the recipient's mind is what was originally in the communicator's understanding, and also to motivate the communicator's further participation in the interaction, there has to be feedback.

During feedback, the recipient will communicate with the original transmitter of information in order to receive clarification, or to take action.

The diagram below illustrates the essential elements of the communication cycle.

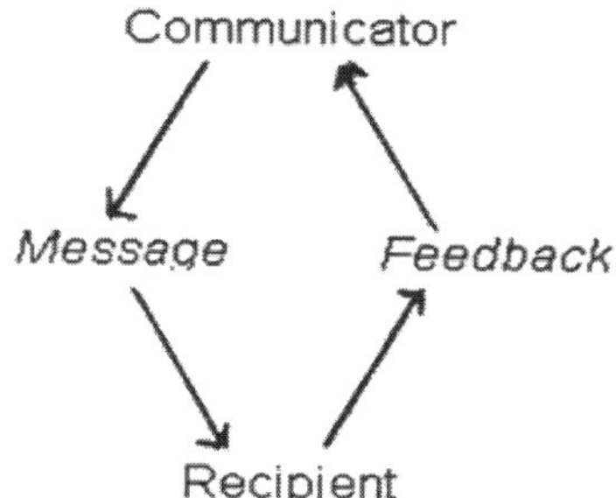

Figure 2: *Elements of the Communication Cycle*

A breakdown in any one of these elements will cause the breakdown of the entire cycle and communication will sooner or later come to a halt. This should be borne in mind when planning, and later when maintaining, the flow of information through a system. There needs to be a clearly defined flow of information with defined inputs, outputs and feedback. Suppliers of information need to know that the products of their labour are being put to good use, or their supply will dry up.

Structure and flow of veterinary information systems will be dealt with in greater detail later, suffice it to say at this stage that information is usually garnered from farmers by some kind of field worker, who will pass it up a chain to a central computer system, from where feedback will be transmitted back to the field.

Managing the System

Provision must be made for the position of system manager to run the system. Such a person will have to carry out the following functions:

- *Monitor data flow* into the system - and take follow-up action when flow slows down, or when inflow becomes abnormally large. Both could indicate abnormalities in the system which need to be remeded. A "drying-up" of data coming in could indicate a lack of motivation of field staff for a number of reasons (including poor feedback), or logistical problems among data gathering staff in the field. A "surge" of flow could mean that there is an upswing in disease incidence, or livestock population, or rampant data fabrication, or that the system had been functioning suboptimally before the surge, and it is now coming to an equilibrium.

- *Check data quality* - various parameters need checking. They vary from simple spellings to handwriting, looking for fabricated data, checking internal logic (for example, when a veterinarian reports zero deaths in an area, but in the same report describes post-mortem lesions). This might also mean cross-checking with data from other sources, providing direct feedback to suppliers of data and requesting them to verify what has already been submitted. The manager must be motivated enough - and have enough "clout" in the hierarchy - to take action when deficiencies come to light.
- Carrying out *data analysis* and ensuring that analysed information reaches decision-makers. This is an ongoing function, and disease data will need ongoing temporal and spatial analysis to look for trends in terms of increasing or decreasing incidence, spread of disease, etc. Vaccination data will need following-up to assess percentage coverage in vaccination campaigns. Again, the manager must not just be an observer. Disturbing trends demand remedial action, and no manager can afford to be a spectator.
- *Ensuring feedback* to the field. Field staff need to receive reports indicating the outcomes of trends in data, and what action is being taken to respond to these trends. This will encourage them to continue with their work, as they will see that what they do is constructive and helpful. It will also give them the opportunity to respond, and voice their opinions as to whether analyses made are reasonable, and whether actions taken are justifiable. In short, it gives them a stake in the system. At the same time, feedback needs to be given to livestock owners to encourage their further co-operation with field staff. This could take the form of information dissemination through pamphlets, posters or farmers' days. Once again, it effectively gives them a stake in the system, as well. It is the job of the information manager to ensure that such feedback is made, and that it reaches grassroot levels.

At country level, it is the *national epidemiologist* who is best placed to deal with overall information management.

In large countries, some of his functions could be delegated to lower levels, but the overall responsibility for maintaining the integrity of the system should obviously - for simple reasons of accountability - be in the hands of one person.

Performance Indicators - Measuring Efficacy of Surveillance

Various parameters can be used to measure the efficacy of surveillance. These should be agreed upon during system formulation, and reviewed from time to time.

Such parameters should be monitored by the epidemiologist/information manager on a regular basis. Following trends every month is probably the best, and the epidemiologist should work out a simple monthly schedule to cater for routine activities such as data checking, reporting and analysis, and monitoring performance indicators. Examples of indicators that could be used are

- number of reports submitted/1000 head of livestock/district/month
- number of individual livestock inspections/staff member/month
- percentage of observed disease incidents for which laboratory samples were submitted
- percentage of suspected cases actually confirmed for any particular disease
- time lag from sample submission to final laboratory diagnosis.

The Joint FAO/IAEA Division and EMPRES has proposed guidelines for use in Rinderpest surveillance for the GREP; these guidelines could be adapted to many situations. The information manager will need to establis the set of performance indicators best suited to his situation, and ensure that these are acted upon when the situation so demands.

Setting the Goals; Determining Needs and Outputs

To some extent, the outputs derived from an information system will determine what sort of inputs are required. In other words, when designing an information system, the first thing that must happen is that the system's future users must discuss their needs with a system designer. What sort of outputs are needed? To quote a trivial example, if we would like to know each month which diseases occurred, which species were affected, and how many animals were affected, it presupposes that the basic data will contain details such as the animal species affected, how many affected, the reporting date and the name of the disease suspected.

A good starting point would be to look at any existing manual system (often regular "monthly returns" that come to a supervisor and gather dust), and ascertain the inputs and outputs of that system. The new system in the process of conception might be able to build, at least to some extent, on what is already established. It will also show very clearly what kind of information is not used, and what could be improved upon to make it more useful.

Computerisation

As mentioned above, computers have become smaller, cheaper, user-friendly, more readily available and more robust. They can be taken into all sorts of environments and programmes are available that will perform a great variety of functions. But just as horses are more suited to certain terrain than the motor car, so we must remember that computers, for all of their hi-tech functionality, simply can't do everything. Computerisation, put bluntly, is not a panacea.

What it Will Do, What it Won't Do?

Computers will not replace good personnel. They do not reach far-off stock owners and collect data. Trite though it may seem, having a computer doth not a system make. The system design must, first and foremost, incorporate people and their abilities, and make provision for extensive training in the use, completion and submission of data questionnaires. In the ordinary course of events, input questionnaires should never be completed by farmers and by those unschooled in their use; only by properly trained personnel. The first principle of data recording is, and always will be, "garbage in - garbage out".

Computers will not check data for you. They will not detect fabricated data, problems with data logic, nor will they improve poor handwriting on data forms. Such work must be done at field level by the most senior staff member available before forms are dispatched to the computer centre for typing-in.

Computers cannot replace the epidemiologist. Computers are simply data storage and retrieval devices. They cannot make judgements on data quality, notice disease trends, contact field veterinarians to ascertain their feelings on the current disease situation, or recommend a course of action. While the foregoing may seem glaringly obvious, it is all too often forgotten by those who want to create the "perfect system" without realising the crippling limitations of computer technology.

Computers are not a short-cut to easy data storage for all types of variables, either. They can be used for storing relatively simple numeric and non-numeric variables, and should only be used for this purpose. The examples given above of computerising livestock movements and storing trade routes are very fitting in the veterinary context. Another "variable" that must not be overlooked is a very simple one - the "gut feeling" of the field veterinarian. The field vet is in touch with farmers and field staff, he knows the conditions of the pasture, the livestock and the local market; he understands the way diseases behave in his area where he may have worked for some years. To computerise this? Impossible - which is why regular personal communication with headquarters is always important.

Computers also do not write interesting reports and do imaginative analyses, nor do they give constant encouragement to field staff. They may store data, but making sense of what is stored is, in the end, a human function. To calculate is computing, to interpret is human.

The centrepiece of any information system is not the computer, or the programmes installed on it: it is the people who run the system, and most particularly the epidemiologist who stands at the centre and directs operations. Without a dedicated, enthusiastic and "wide-awake" epidemiologist, the system will crash as surely as it will when the computer suffers a power failure.

National Computer Systems

There is often a tendency to try to over-computerise, which can be fatal. In other words, some would have a computer in every district veterinary office - after all, what could be easier than local data input and simple electronic transfer to the central database. This concept has several major shortcomings:

- Who will be responsible for data input? And who will check it before it is transmitted upwards to the central computer? A busy veterinarian will certainly not type in large numbers of data forms reaching his office, and it will be left to a clerk. What of the work load already carried by the clerk? Or will one need to employ an army of data input clerks to cope with each district office? Data input is a very monotonous but highly demanding job that requires high standards of accuracy. Ordinary office clerks are usually unable to cope with such work, and when it is imposed on them, the results are often disastrous - slapdash and shoddy, with a wide variation in data quality between various staff members. In addition, if good data entry clerks are employed, their capabilities are usually such that they could easily cope with work from a number of districts simultaneously.
- Who will maintain the system? A huge network of computers strung out across the countryside means regular breakdowns,

staff having problems with software, power failures, etc. This implies a number of technicians to keep order, and ensure a smooth flow of information.

- More importantly, who will pay? The more computers, the more connections, the greater the cost - the cost of maintaining the large number of machines, regular upgades, and paying for the connectivity which will, obviously, be supplied by the national 'phone company.

Large numbers of staff with sophisticated, sprawling computer networks certainly have their place in developed countries, but in developing countries the secret is rather to start too small than too big. If one takes a convenient number of 20 to 25 veterinary districts to one computer, then it is quite practical to say that in large countries, a small computer unit at regional/provincial level could handle the inputs from those administrative divisions and than pass the data on to the central unit.

In very small countries, a single central data input and processing unit would be sufficient. Admittedly, one would have to trust to national postal system to get data questionnaires to the input unit, but that is often preferable to trusting a strung-out computer network with all of its associated complications. The easiest way to handle this would be for each station to post all of its inputs on a regular basis (fortnightly; monthly) to its computer unit, and to keep duplicates (carbon copies are cheapest) of all questionnaires in reserve.

Information Backup

Any information system must have a data backup capacity. Backups are kept in a variety of ways:

- duplicate copies of completed questionnaires at field offices.
- original copies of completed questionnaires, sorted according to district and month at the data input centre.
- electronic backups of data that has been stored on computer. Data may be backed up on tapes, diskettes, or compact disk, but *it must be backed up.* Some computer systems make provision for automatic daily backups, and these can be timed to take place after hours. In the absence of an automatic backup facility, data should be backed up at least once a week, so that if a computer crash occurs, the amount of information that has to be re-entered is minimised.

When making backups, especially onto tapes or CDs, the "grandfather-father-son" principle should be followed. If, for instance, the first backup is made on CD no. 1, the next on CD no.2 and the third on CD no. 3, then to make the fourth backup, the user will revert to CD no. 1. This ensures that the most recent backup is still available and unscathed should a crash actually occur during the backup process.

Running a System Without a Computer

It may seem unusual to include this topic in a manual written in the computer age, but the fact is that information systems can exist without computers, and in some cases, they simply have to. In far-flung areas where electricity supply is erratic or non-existent, in very poor countries or regions, or in small projects run by NGOs, computers may be inappropriate, impractical or simply too expensive.

In such cases, data are stored on large tables called tabulation sheets. Each column in the sheet will contain one of the variables to be recorded, while each row of the table will represent the information contained on a single questionnaire. It is self-evident that questionnaires will have to be short and simple, and data volumes low! Data entry staff will transfer data from the completed questionnaires to the tabulation sheet by hand.

Questionnaire Design

Questionnaires can be used in a number of ways. Firstly, they can be used as a rough guide to a discussion, and as concrete information arises from the discussion, it can be entered onto the questionnaire. Alternatively, notes can be kept of interviews (whether person-to-person or group interviews) and pertinent information can be transferred to the questionnaire at a later stage. Or, and this is most usual, the questionnaire is used directly during a formal interview, and its structure strictly followed, point for point.

No matter which way a questionnaire is used, its design is of great importance to the success of information gathering, and a number of important principles apply:

Content : This refers to the actual variables to be recorded. They will have been decided during the design stage of the information system, and must be kept to a minimum.

Time: The length of the interview (which is directly determined by the number of variables to be recorded) must be short. 15 minutes is a practical guideline, shorter is better. When doing complex surveys (eg. Rapid Appraisals), interviews may become longer, but under those circumstances, anything longer than one hour is excessive.

User-friendly: The questionnaire layout must be clear and logical so that the interviewer follows a logical sequence down the page from start to finish. Likewise, it must provide a logical sequence for the data entry clerk to follow when transferring data from the questionnaire to the database. It must also be clear and legible. Certain parts of the questionnaire may contain comments or information not intended for the database - these should be clearly marked.

Self-contained: The form must be self-contained in that all necessary information is contained on it - district, date, details of interviewer, name of place, disease information etc.

Coding: Where possible, information should be coded on the form to simplify and accelerate data input. If the code for a particular district is a set of letters, the field worker should enter his district's letters on the form - or they could be pre-entered before a field visit is undertaken.

Species, such as "Bovine" could be ticked off in an appropriate "check box" on the form. Nonetheless, on-form coding should be approached with care. If check boxes for every conceivable alternative of every variable are provide on the form, it will become complex and possibly difficult to read and complete - perhaps leading to the wrong boxes being checked, or even omitted.

Dealing with purely "veterinary" information means that it is often best given in narrative form and then encoded at input level - for example, issues such as clinical signs or post-mortem lesions. Giving a few alternative signs on a form will lead to all syndromes reported having very similar appearances. Such a form of information collection could also have the effect of being a set of leading questions, in which farmers are lead to describe a particular condition.

Consider the possible effects of the following interview:

> *"Did your animals show discharges and diarrhoea before death?"*

As opposed to:

> *"Please describe all abnormal signs you noticed before your animals died."*

The first question is clearly leading, and if animals did not show clinical signs within the narrow range given, confusion will result. Either the stock owner will assent to the few symptoms given, even if they were not seen (after all, his animals *did* die) or he may be reluctant to co-operate further. The interviewer will simply record what he is told, blissfully unaware of whether it is valid or not, and garbage will enter the system.

The second question allows a full description of what really happened, and the interviewer will get closer to the truth. Instead of being forced to fit all diseases seen into what will perforce be narrow sets of clinical signs, it will be possible to identify a wider range of diseases. It will thus be possible to get early warning of new - and hitherto unknown - diseases in the area.

Clinical signs and post-mortem lesions are best left as open as possible so that farmers have a free reign to describe exactly what they saw, and are not limited by a few choices on a questionnaire. In such cases, encoding can be done at the level of computerisation, where a full list of codes to cover every symptom and lesion can be kept, and used during input.

Presentation

Issues such as paper size and quality, clarity of printing and size of spaces for recording answers deserve careful consideration. A smart, simple questionnaire will further ensure good quality information. Shoddy, overcrowded and complex forms will be completed in a shoddy manner.

Questionnaires should be so designed as to accommodate all diseases, not custom-made to suit just one. Expecting field workers to carry a variety of questionnaires, each for use with a different disease is wasteful of resources and will only cause confusion. (The same, incidentally, goes for the database - creating a separate information system for each disease is senseless. The entire system should be broad-based enough and robust enough to cope with any eventuality. Anything less than this reflects poor system design).

A few specimen questionnaires are given at the end of this manual as examples.

Start-up Hints

Using questionnaires for the first time means a few things. First of all, an instruction manual should be prepared, explaining the questionnaire in greater detail. The aims of the questionnaire should be summarised. What is required under each data item (in logical order) should be explained.

Once this has been done, a number of the field workers who will be using the questionnaire should be trained in its use.

A briefing on the contents of the questionnaire and how best to put the questions should be given first, and then the workers should use role-playing techniques to test their abilities among themselves. A field test is then carried out, with a senior staff member (preferably the questionnaire designer) accompanying each field worker for the first interview. Each worker is then allowed to do a few further interviews on his own.

After this exercise, the response of the field workers is evaluated, and the first data are examined in order to determine shortcomings with the questionnaire and modify it if necessary.

Databases

Databases, in their most general sense, are simply a means of storing data - whether it be on a computer or on a tabulation sheet, in a card index system or in a ledger. In recent times, however, the word "database" has almost become synonymous with "computer". Computers were, in fact originally conceived as high-speed data storage and retrieval mechanisms.

As computers are evolving at such a frightening speed, no attempt will be made to suggest specifications in this manual. Prospective users should consult with a number of vendors in order to gauge trends and make intelligent purchases.

Functions

Database software is generally able to store most types of data - whether as numeric or non-numeric variables, sort data, and retrieve specific items or subsets of data in response to user queries. Data files, or specific subsets, can also be exported into spreadsheet programmes for graphic analysis, and where data are georeferenced, can be introduced into GIS software to be visualised as maps.

As mentioned earlier, it is important to distinguish between data on questionnaires that are destined for computerisation and those that are not (eg. specific comments from field staff to their supervising veterinarian, or information cattle movements that needs to be followed up), and also to take note of the fact that certain information, while it must be computerised, will need action before it even reaches the computer (eg. a suspicion of a new disease outbreak).

Software

Various database programmes are currently available, and while these may change in the future, they can probably (at the time of writing) be divided into three broad categories:

- Software for small business (and therefore for relatively small datasets) - eg. Microsoft Access.
- "Middleweight" software for medium-sized databases of all types - eg. Microsoft Visual FoxPro, Borland Visual dBase and Borland Paradox.

- Software for large-scale databases (often used by banks, large companies, government ministries, international organisations) - Oracle, Sybase.

No blanket recommendation can be made here regarding the "best" software to use, and the above can be taken as examples only. Software requirements will vary from system to system, and obviously according to data volumes collected. What is important is ensure that the hardware (ie. the computer) has the capacity to run the chosen software, and the system overall can cope with the input volume and the outputs required. What also needs mentioning is that there are some more or less custom-made software packages available for epidemiological data storage and analysis. Some of these packages are old and most have very limited capacities. Database software is normally good for storage, quick sorting and retrieval of data, and small statistical manipulations (eg. range, mean, std deviation) are sometimes possible. In order to carry out advanced data analysis, spreadsheet packages (eg. MS Excel, Corel Quattro-Pro) and statistics packages are needed (eg. StatGraphics, Kwikstat, Epi-Info). For spatial analysis, one moves into the advanced world of Geographic Information Systems (eg. Arc View, MapInfo).

File Structures

Information is stored on computer in entities known as "files". Naming of files, and what goes into them (ie. their structure) is the prerogative of the user. Database files are structured, with each observation (in our case, the equivalent of a questionnaire) recorded in the file as a "record". The variables in each record (eg. locality name, animal species, number sick, etc) are known as "fields". A group of records with a homogenous structure, and stored together, is known as a table. There are different types of variables, and the software will handle each type differently - for example, a place name would be stored as a "character" variable, and the programme would be able, for example, to sort them alphabetically. The number of animals reported dead in an outbreak would be stored as a "numeric" variable, and calculations could be performed upon it. Variables and their types must be defined during database design, and must exactly follow the structure of the questionnaire. As this is not a manual on database software, there will be no further discussion of software at this stage. Enough information on databases is available elsewhere.

Coding

As part of database planning, it is essential to choose a set of codes for each variable before the database is launched. For example, one disease may have many names, such as blackleg, black quarter, or quarter evil. Which of these to use in the database? It would be preferable to type in a standard code for the disease, such as BQ (or some other suitable, but recognisable, code). The data entry clerk could be supplied with a "look-up table" (which may be programmed into the computer, or kept in a separate manual), look up one of the synonyms, and enter the correct code. Likewise, when it comes to symptoms, "drooling" and "salivation" might be encoded as SALIV.

Encoding means that information can easily be retrieved. When one wants to enquire about black quarter, it would not be necessary to ask separately about each of the three synonyms in order to get a full picture of the disease - one query using the code "BQ" would immediately render all the information available to the enquirer.

Data input staff soon become very familiar with the most common codes and after initial "teething" periods, will be able to enter the correct codes for most variables - districts, species, diseases, clinical signs, etc, almost without thinking.

Ease of Input

Making inputs into the database must be a simple operation. The "user interface" through which the input staff have to work needs to be simple and friendly. Typing into a template and having on-line assistance available is a great help, and the order of input of variables must follow the order of the questionnaire. Having an easy-to-read manual written to help data entry clerks is ideal, and of course, training is essential. Data entry staff must also be chosen for their speed and accuracy, and their ability to cope with monotonous, repetitive work.

Querying the Database

Most queries will be fairly simple, and most modern software has "query builders" built in which the user can employ in analysing data. Queries usually take the form of something like "list all the foci of pasteurellosis in bovines in district *x* for the month of June" or "calculate the sum of all CBPP cases diagnosed during the year."

Exactly how to formulate queries will be described in the software manual, and on-line help and hints are usually available within programmes. It is important that epidemiologists perform a standard set of queries very regularly (say each month) for reporting purposes. Having a set of standard outputs and analyses keeps field staff up and management alike up to date with the disease situation in the country, and helps foster confidence in the system.

Data Quality Control

Although data control has been mentioned earlier, it deserves further treatment at this point as a separate entity.

Data quality control is an integral part of information management. As has been made clear elsewhere in this manual, it is a fatal mistake to assume that all data entering a system are good data.

Data move from the field to the district office to database input. The more checks are conducted before input, the better. If a problem is detected while a piece of information is still relatively near to its source, it can be followed up and corrected with relative ease, the further data move from their origin, the more difficult - and costly - corrections become.

Checking levels and what is checked are as follows:

In the Field: Careful questioning of the farmer to capture a true reflection of epidemiological information. Leading questions should be avoided. If information comes from farmer recall, it may be worthwhile to cross-check information with other family members or in contact farmers.

At the District Office: Completed questionnaires are evaluated for legibility, correctness (eg. place names, code usage) accuracy and internal logic. What is written must, in other words be clear, neat and make sense. Where a query arises, the district supervisor (preferably a veterinarian) must first contact the interviewer concerned to clarify the issue with him. If necessary, and if possible, a return should be made to the original data source (the farmer) to follow up. Not only is it easier (nearer) to do this while still at field level; it is also possible to recapture information while it is still within reasonable recall and important details are not yet forgotten.

At the Epidemiology Unit: The data entry clerks will detect - and complain about - poor handwriting. The epidemiologist will further do spot checks on individual questionnaires before data entry, and also cross-check data entered onto the database with the questionnaires from which the data came on a random basis.

Data input staff will need good training and careful monitoring. It essential that data typists do not sit in front of computers for extended periods, as this leads to physical tiredness, eye and mental fatigue and a lack of concentration. Where possible, data entry should be interspersed with other tasks, such as the sorting and filing of questionnaires, doing data backups, sending enquiries to the field about data quality, etc.

A very important task of the epidemiologist, mentioned only in passing thus far, is the detection of fabricated data. It is a painful truth that some field workers will not always visit each place on their visiting programmes, but may simply sit at home and complete questionnaires in their easy chairs. Discovering such data is not easy, but a few pointers might help. It is helpful to carry a preliminary analysis of data and look for tendencies. First, simply view recently entered data in its "raw" form in the table, and then carry out a few simple statistical procedures, such as calculation of range, mean, standard deviation, construct histograms, view data spatially with a GIS. Look for the following:

- repetition of the same, or similar values (eg. herd sizes, or numbers of animals affected by a particular disease).
- variables having a small standard deviation, with many values clustered around the mean.
- a very homogenous disease/clinical signs pattern in a particular area.
- many values ending in zeros or fives.
- herd size distributions that are vastly different in adjacent and very similar areas.
- many observations of herd inspections with little or ancillary information given on the questionnaires (eg. questions on clinical signs or pasture conditions may be unanswered).
- details given for a visit to a particular place may be identical to another visit to the same place six months previously, ie. the previous visit's questionnaire was copied.
- a particular disease being widespread in one worker's area while being absent from an adjacent area where prevailing conditions are very similar.

All of the above are reasons to be suspicious about data. Where there is a pattern of repetition in a particular worker's information, then his data are suspect. Where there are vast disparities between data of two adjacent field worker's areas, it will be difficult to say exactly which of the two is at fault. In general, however, it may be necessary to send an independent team into the field to conduct random inspections in the areas from which devious data have come as a verification study. Results can be compared afterward. Where it is obvious that a field worker has fabricated data, severe disciplinary steps must be taken.

Needless to say, field management of veterinary staff remains an important aspect of basic management, not just data management. Staff must work according to fixed programmes, and spot checks must be made by supervisors from time to time to ensure that they are actually "on programme."

Errors need not only be the result of fabrication, though. Biases in data can arise for other reasons, and epidemiologists must be aware of these. Routine surveillance is very often not randomised, but pre-programmed, and anomalies may arise:

- field workers might use motor vehicles and not always reach more remote areas. This may confine field work to areas near main roads where richer farmers often live. Such farmers usually have larger herds and flocks, and use more medicines and vaccines - with a correspondingly lower disease incidence.
- field workers are usually male. For various reasons associated with personal prejudice or cultural taboos, they will not interview women, and so gain inaccurate data on animal species with which women usually work, eg. poultry and small ruminants.

Even where surveys are randomised, errors will creep in. If livestock numbers are incorrectly estimated, serum sampling tubes may be too few, resulting in unrealistically small sample sizes. Sampling animals of unknown vaccination history may result in the detection of vaccine titres during sero-surveillance. Farmers may lie about what diseases their animals have had.

Data input staff may quite literally have a bad day and miss the typing-in of a batch of data forms - or forms may get lost in the post, leaving a "hole" in the database.

Lists of what can go wrong are endless and very depressing, but can be minimised through:

- thorough staff training (at field and headquarters level)
- creating strong farmer awareness and gaining their co-operation
- good planning of data collection, routine surveillance and special surveys
- enforcing strong discipline amongst staff
- having a vigilant epidemiologist

Verification

Giving feedback to the field in the form of analysed data regular reports is a good way of seeing whether trends registered in the database are a good reflection of trends on the ground. Another tactic - and this can be expensive - is to send each field veterinarian a monthly printout of all data entered from his district during the preceding month for him to inspect, correct and return to the epidemiology unit.

In this way, records that were incorrectly entered can be corrected, and should the veterinarian wish to update some information on particular incident, he has the opportunity to do so. If this is too expensive, sending each veterinarian a six-monthly summary of his district's data may be a more viable option.

Visits by epidemiology unit staff to the field are an indispensable means of maintaining contact, and an opportunity for on-the-spot validation studies.

Feedback

The importance of feedback has thus far been mentioned several times. It maintains the chain of communication and ensures interest on the part of field staff. The question is, how?

Regular reports giving summaries of the disease status in various parts of the country are probably ideal. Such reports must be clear and interesting, well illustrated with graphs, maps and tables. Veterinarians must be encouraged to comment on such reports, write "letters to the editor" and contribute short articles on interesting cases. A lively rapport between the epidemiology unit and the field is certain to keep the information system alive, while at the same time ensuring that all field staff are kept well informed. Part and parcel of this communication will be regular meetings between district vets and their field staff to discuss the contents of such regular reports. It is of the utmost importance that this feedback reach the people who collect the data on the ground. They must be made to feel part of a team.

It is a good idea, as mentioned in the previous section to send each district - or even, where practically possible, each staff member a short summary of their reporting every six months. A short table giving disease totals together with one or two illustrative graphs is sufficient for each person to know that his data area received and appreciated. It also gives each person a "snapshot" of the situation in his area.

The Role of GIS

GIS stands for Geographic Information System. A GIS is an automated (ie. computerised) system for the input, storage, analysis and output (viewing) of spatial information. Various software packages have been developed for the visualisation of geographically referenced data, all them resource-hungry, all of them expensive, and, for the epidemiologist, usually indispensable. In fact, running a veterinary information system without some kind of a mapping system - even a manual one - would be akin to trying to run a motor car without a speedometre or a fuel gauge.

The main advantage of GIS software is not just that the user is enabled to see how a disease is distributed geographically, but also that an animal disease can be viewed against other information - for example, rainfall maps, vegetation maps, rivers, and so on. The disease presence can then be related to other factors and more easily appreciated visually.

Georeferencing Data

Using disease data within a GIS means that every observation of disease recorded in the disease database must be georeferenced (in other words, must be accompanied by latitude and longitude values or some other grid reference recognised by the software).

There are two main ways of handling this, each of which has advantages and disadvantages. One is to add georeferences to each report (ie. each questionnaire must have a place where georeferences can be filled in). The other method is to have a "master file" built into the computer database where the computer (automatically) or the input clerk (manually) can "look up" the latitude/longitude values and add them to the records.

Georeferences on the Questionnaire

This can be done in a number of ways. Each field worker can be given a standard list of all georeferences for the places in his area of work, and can add to the data forms as he goes along, or he could look them up on his return to his district base and add them there. Alternatively, someone else at the district office could fill them in once the forms have been handed in, which might be preferable, as many field workers (especially those with lower educational levels) might not understand the principles involved, or might make too many errors. The district veterinarian would have to check the co-ordinates before sending the forms away for computerisation, but obviously would not be able to check everything in detail.

Getting Georeferences from a Computer Table

This would entail having a database table containing the names and geographic co-ordinates of every place in the country that could conceivably be visited by field staff. From that point, things could happen in one of two ways:

The data entry clerk would enter the details from the lookup table into the main database each time an observation is entered. This poses some problems. Should the place name perchance not be in the lookup table, there would be no easy way to quickly ascertain the co-ordinates. The same would apply if the place name were misspelled on the questionnaire (hopefully a thorough check at district level would have obviated that possibility), or if place names had changed (as they sometimes do in traditional parts of Africa, for example). To make matters worse, one sometimes finds places in the same country and even the same area with the same names, giving rise to the question of which one is referred to on the data form.

The other possibility is to have software which scans the data entries and the lookup table systematically, and then enters the georeferences automatically after cross-matching place names in the disease data table with place names in the lookup table or master file. The same problems noted above for manual georeference entry would apply for such automatic entry. Both concepts of georeference entry into the database - either at field level or at the computer level, are fraught with problems, and there are no easy solutions. Each country and system designer would have to cope with the issue in the way best suited to their circumstances.

Using the GIS

Nothing much will be said about use of GIS software, and the reader is advised to consult relevant literature on the subject. The prospective user must consult thoroughly before purchasing such software, bearing in mind the following points:

- the software must be compatible with the computer system in use. GIS programmes are usually resource-hungry, requiring much disk space, operating memory and fast processors. It may be necessary to upgrade an older computer system in order to accommodate modern GIS programmes.
- the GIS software must be compatible with the database system in use. The GIS treats different kinds of data as "layers" and essentially, disease data would be imported into the mapping system as a layer. It must first be ascertained whether the database format in use in the epidemiology unit can be "read" by the GIS; the next is to ensure that the manner in which the georeferences are stored can be read by the GIS. It may be necessary, for example, to convert degrees, minutes and seconds into decimal degrees (where, for example, 30° 30' 30" would be read as 30.5 degrees) before importing the database file into the GIS, or the GIS may do it automatically.

- there must be ready maintenance for the GIS software within easy reach, and it should be easy to acquire important layers (roads, weather data, etc) locally. Usually, GIS programmes when purchased, come with a world map containing little more than country borders and the grid references for capital cities. More details have to be purchased, begged, borrowed or stolen elsewhere.
- training and other support should also be available. GIS software, while having many advantages, and being important in good epidemiological practice, is notoriously difficult to use. The importance of good training cannot be over-emphasised.

Motivating and Training Field Staff

Field staff must be carefully chosen and well-trained. Farmers are usually willing to part with information when they feel that field staff are intelligent, trustworthy and can offer something in return. Field staff should be able to offer advice on matters related to animal health, and should also be able to carry out "first aid" treatments should they encounter sick animals during the course of their rounds. Treatments need not necessarily be for free, but the service should be available. They should also be well-trained in questionnaire usage.

The ability to handle animals well, take blood samples competently and to interact with people are also essential prerequisites.

The same principles apply when using "outsiders" to collect information. There is a strong tendency to make use of Community Animal Health Workers or "barefoot vets" from the private sector as gatherers of information in the field. They are interviewed periodically and all disease information is transferred to questionnaires for computerisation. It should not necessarily be assumed that such people have all the training that they need. They too will have to be trained in certain basics before being used as trusted sources of information.

Basic Training for Livestock Inspectors or Community Animal Health Workers

The kind of subjects that might be covered in such training (which must have a strong practical component) would include:

- The very basic principles of anatomy and physiology of the domestic animals in the area.
- Principles of nutrition and pasture ecology.
- Animal diseases of local importance: clinical and post mortem signs, epidemiology, prevention, treatment.

- Applying first aid, the use of basic veterinary medicines (wound treatments, dips, anthelmintics, antibiotics, trypanocides, babesiacides, vaccines, care and storage of medicines and vaccines, use and care of syringes, etc).
- The basic principles of sero-surveillance campaigns - how to draw blood, store sera, etc.
- Questionnaire usage, information recording and interview technique as appropriate.
- All teaching should be illustrated with colour slides and/or posters, and adequate practical work under local conditions is all-important.

Training should not be conducted such that all subjects are dealt with in one course. The course should be divided into modules, between which the workers will return to the field and apply their knowledge in practice. The teaching should be reinforced with frequent revisions. Evaluations should be practical rather than in the form of written examinations. While the training above is aimed at lay persons, it should not be forgotten that veterinarians will also need training is questionnaire use.

Logistics

Backing up field staff with sufficient supplies is basic to their functioning. Field staff who are undersupplied with their necessities will become frustrated and discouraged and will not function. The following issues must be considered:

- regular pay and access to food and clothing supplies if the staff are on the official payroll.
- transport in the field. Vehicles and motorcycles will require fuel, regular servicing and a supply of spare parts.
- protective clothing. There should be a regular issue of overalls and appropriate footwear.
- camping equipment should be available, especially for the rainy season.
- a plentiful supply of questionnaires, carbon paper (to make duplicate copies), pens, pencils and clipboards is essential.
- basic medicine kits with a few essential stock remedies, needles and syringes.
- other supplies, such as serum collection tubes, labels, markers, ear-tags, vaccines, cold boxes, and so on, must be readily available at the appropriate times.

Awareness Creation Among Decision-makers

Many veterinarians in senior positions are currently not familiar with modern information systems development, and may have false expectations of information systems. They may even be suspicious of them, or simply discount them.

There are those who treat computers with contempt after having had one or two bad experiences with them, or who simply do not understand them. Others, on the other hand, may be over-enthusiastic, and expect too much. Some may think that installing a computer and some database software will create an information avalanche that will enable easy decision-making overnight. The latter group will suffer the most disillusionment when they discover that implementing an information system is a slow and painstaking process, and that the first information to come out of it is untrustworthy.

Having a computer system will not:

- Automatically improve information collection or quality. It may provide an impetus in this direction, but will not do so of itself.
- Provide instant disease status information within a few days of installation. Getting information into a system is a long process, and upgrading the quality of information to make it reliable, is an even longer process. It will take one to two years or perhaps longer before outputs are intelligible and usable.
- Replace a good epidemiologist and common sense.

Having a computer system will:

- Take time to plan, install and implement.
- Have a lot of teething problems.
- Bring about cost savings in some areas. Blanket disease control campaigns are the norm in many countries. Having good disease data will show what diseases are present where, and what their incidence levels are. Control mechanisms will then be implemented more selectively and only in the areas where needed. It is a fact that, until recently, many African countries were still carrying out huge annual vaccination campaigns against rinderpest unnecessarily, simply because they had not quantified the threat and correctly identified the areas where it existed.
- Enable the monitoring of progress with disease control efforts, provided that consistent, accurate reliable data are being collected.

- Require a lot more headquarters-field liaison, and giving large quantities of feedback and encouragement.
- Cause management staff to appreciate field personnel a lot more and build valuable relationships and *esprit de corps.*
- Make decision making on disease management much easier - once the system is properly running.

Computer-analysed data form part of the information that goes into veterinary decision-making processes, but only a part. All planning must be done against a background of knowledge of staff capabilities, budgetary constraints; cultural values and traditions, and prevailing government policy. Decisions must be holistic and take all of reality into account, and not just the slice of it that comes from a computer.

Getting a System Started

Leaving this section until this late in the manual has been deliberate. Having an overview of what a veterinary surveillance system will entail is essential before trying to design and implement one. Some ground has already been covered in previous pages, but some aspects still need discussion.

- Agreeing on the need for systematic data collection and analysis is an essential first step, that will require intense discussion between all stakeholders - management, the epidemiologist, field staff and laboratory staff.
- Designing the system, taking into account any system that already exists, is the next step. Crucial decisions must be made regarding data sources, inputs and outputs.
- Database design is the next step, together with software customisation and questionnaire design. Several questionnaires may need to be designed according to information sources-visual surveillance, abattoir surveillance, and so on.
- The next stage is field testing. It would be most practical to do a few pilot runs in two or three selected areas, completing questionnaires, following the information flow into the computer, and then doing some trial analyses. Shortcomings can be identified and corrected.
- Then all veterinarians will have to be informed and trained in implementing the system, and they in turn will need to train their field staff.
- The system is then implemented. Data flow and quality will have to be intensely and carefully monitored in order to make

as many improvements as possible. It is important that both management and field staff be informed of what is happening every step of the way. If left in the dark, suspicions will develop, and enthusiasm and support will wane.

- All in all, it will take one to two years from conception to implementation. Allowing for some hiccups along the way, another year or two for the system to mature enough to produce "research quality" data.
- Bear in mind that no information system is ever static. It will continue to evolve and grow as needs change and users become more sophisticated.

Using Surveillance Data as a Management Tool

It is not the intention here to write a textbook on veterinary epidemiology, but rather to give few pointers. From the point of view of an epidemiologist, having a strong flow of "analysable" data can be very exciting. The trick is to make good use of it. First and foremost, it must be remembered that an information system is a management tool. Using information from a system for research and retrospective studies may be exciting, but day-to-day management is the aim.

Regular preparation of month-by-month incidence graphs of the main diseases can be very informative. Doing such work for one or two years may reveal important seasonal tendencies, particularly as far as vector-borne diseases are concerned, but possibly with others, as well. This may give good indicators as the planning of official strategic controls or regular reminders to farmers.

Longitudinal studies will also be of value in monitoring progress with eradication efforts, deciding when best to enter an OIE pathway, or to change from visual surveillance for a particular disease to sero-surveillance or abattoir surveillance.

Spatial analyses using a GIS are extremely useful for monitoring disease spread, and for noting the association a disease may have with weather conditions, ecological zones or geographic features such as rivers or roads. Areas under particular threat can be noticed early in the development of a disease, enabling the more rational deployment of resources. Planning for staff distribution, transport arrangements and vaccine purchases is facilitated.

Long-term studies of disease data will also reveal cyclical tendencies, association with long-term climate changes, and so on. It may even be possible to build reliable models of some diseases which enable predictions of spread to be made with greater accuracy. An important fact about disease information systems is the ability to link them with economic data, and make inferences regarding economic impacts of diseases and the costs and benefits of control measures. Computer software can then be harnessed for building reliable "what-if?" scenarios and allowing an intelligent choice of various control options to be made.

FAO Involvement in Information and Surveillance Systems Development for Transboundary Animal Diseases

The preparation of this manual is testimony of FAO's commitment to assist developing countries with development of their own early warning systems. Via the EMPRES programme, FAO is involved at national, regional and global level with the development of disease early warning systems. The ultimate vision is a global network, linking member countries in an information network that will enable rapid disease reporting, and quick dissemination of information.

This network will be a part of the Global Early Warning System (GEWS) being established by FAO to cover all possible pests, diseases and natural disasters.

EMPRES is currently involved in the development of a three-tiered information system which will gather, process and disseminate information. It is essentially a computerised system, to be known as the Transboundary Animal Disease Information system (TADInfo). This will consist of three different software modules: TADInfo National, TADInfo Regional and TADInfo Global.

For countries lacking properly developed epidemiological software, TADInfo National will be available free of charge. Through the well-established TCP system, FAO will be able to assist with information system development and software installation. TADInfo National is designed to feed information upwards to TADInfo Regional, which will be installed at the level of collaborating regional organisations or projects (such as PARC, SADC or PANAFTOSA) or at regional/subregional FAO offices. Where countries already use their own internally developed software, provision can be made for feeding-in of information from these systems.

Finally, the regional TADInfo modules will feed information to the global module, located in FAO headquarters.

Basically, the functions of the different modules will be:

- at national level: storage and analysis of disease information to facilitate local decision-making.

- at regional level: regional early warning, regional support and co-ordination.
- at global level: risk modelling, trend monitoring and global early warning.

FAO will also take the initiative of organising regional workshops for veterinary epidemiologists to share and disseminate information on disease surveillance.

At the Annual OIE General Session, held in Paris, France in May 1998, FAO was given the mandate, along with OIE and WHO, to build a global information system for disease early warning. This resolution (no. XIII of the 66th General Session) supports an earlier mandate from the 1996 World Food Summit.

The FAO is fully committed to this ideal, and will continue to work towards it via:

- Software (TADInfo) development;
- In-country support in the form of TCPs;
- Regional workshops;

  Interested CVOs and national epidemiologists should contact their nearest FAO office to enquire about the ways in which FAO can assist with the building of national and regional information systems.

Chapter 5

Health Protection and Sanitation Strategies for Cattle

Livestock owners and industry personnel who support their farms are genuinely concerned with the health, wellbeing and productivity of Ontario's cattle. They recognise that disease outbreaks are preventable. They adopt health management practices to prevent the introduction and/or spread of diseases in Ontario's herds. There are very sound economic reasons for disease prevention. Some herd owners spend thousands of dollars each year fighting disease outbreaks. There are many more costs associated with chronic diseases that may go unnoticed. In addition to the costs of health care, valuable livestock and production are also lost. Animal welfare, pride in stockmanship and peace of mind are also major incentives to minimise disease occurrence. Some diseases can be spread through the air. These are difficult to prevent. However, good biosecurity, nutrition, early disease detection and overall health management will help minimise the impact of air-borne diseases.

This Factsheet describes management strategies to prevent the introduction of disease to a farm or control the spread of disease amongst animals within a farm. Although the Factsheet refers specifically to cattle, the general strategies are applicable to other farm livestock. The management strategies that prevent the entry and spread of disease can also be called the "biosecurity plan" for the farm. Every farm should have a biosecurity plan as part of its overall health management strategy.

A section on the control of foreign animal diseases appears at the end of this Factsheet.

Prevent the Introduction of Disease

Management of New Arrivals

Contagious diseases are transmitted directly from an infected animal to an uninfected animal. This is the most common method of disease transmission amongst animals. There are four main strategies for managing the potential introduction of disease when adding animals to the farm.

Maintain a Closed Herd

The first method is not to purchase cattle. For practical reasons there are few truly closed herds in Ontario. Owners would have to strictly adhere to the following requirements:

- use home-grown replacements for maintaining and increasing herd size
- prevent fence-line contacts of their stock with other cattle
- use artificial insemination for breeding and not bring in bulls
- not exhibit at shows
- restrict visitors.

Isolate New Arrivals

Quarantine of incoming animals is ideal. In most herds, minimising contact with the rest of the herd may be the only practical method of isolation. To isolate new arrivals:

- use separate housing, feeding and birthing areas (ideal)
- use separate housing and feeding areas (acceptable)
- prevent contact with other animals (minimum acceptable)
- prevent manure movement from the isolation area to the rest of the herd
- isolate for 21-30 days
- observe and examine for early disease detection
- milk isolated cows last
- test for diseases prior to addition to the main herd.

Know the Source of Purchases and Use Laboratory Testing

Many owners take precautions when purchasing animals. They also use laboratory-testing programs to maintain minimal disease herds or disease-free herd status. To know the health status of herd additions:

- purchase pregnant or virgin heifers to minimise the risk of introducing mastitis

- determine the vaccination and health status of individuals and the herd of origin
- purchase from herds of known health status such as those certified under the Canada Health Accredited Herds program.

The 21-30 day isolation period is ideal for:

- bacterial culture of milk and
- blood testing for specific diseases.

Use Vaccines

Vaccines are commonly used to protect cattle against respiratory disease and abortion. For herd additions, these vaccines may be given during the 21 to 30-day isolation period. Bovine virus diarrhea (BVD) and infectious bovine rhinotracheitis (IBR) have been diagnosed in Ontario herds. Vaccination against these two diseases should be the cornerstone of every herd vaccination program. Consult your veterinarian for specific recommendations on these and other aspects of health management for livestock.

Prevent the Spread of Diseases

Management of Farm Traffic

Bacteria, viruses or other agents of disease are infectious when they are capable of causing infection in exposed animals. Farm visitors wearing boots or clothing freshly contaminated with infectious agents can spread diseases within a farm and among farms. Birds, rodents, pets, people, equipment and vehicles contaminated with manure (or other bodily excretions) are potential disease carriers.

Control Birds

Pigeons, sparrows, starlings and swallows are the most common birds found in and outside barns. They may carry infectious agents on their feet and within their digestive system. To control bird populations:

- plug small and large nesting holes and perches in your barn that are suitable for sparrows and starlings
- screen all openings in natural ventilation dairy barns
- seal off openings into silo roofs
- screen ledges used as nesting sites by pigeons.

Control Rats and Mice

A rat deposits 25,000 droppings and a mouse deposits 17,000 droppings in one year. Even a small population of these rodents may severely contaminate feed supplies. In addition, rodents carry disease agents on their feet and fur, and they destroy millions of dollars worth of feed, supplies and buildings each year. To control rats and mice:

- construct rodent-proof buildings
- eliminate safe hiding places and nesting sites
- remove food and water supplies
- destroy existing populations by poisoning, fumigating, or trapping.

Control People and Pets

People spread contaminated material directly on footwear, hands and clothing. To decrease the spread of contaminants:

- inform herd workers, visitors and truckers of your farm protection methods and insist upon co-operation
- discourage visitors from entering the housing and feeding areas
- post "Do Not Enter" signs with a telephone number for contacting you on livestock buildings and farm entrance gates
- discourage visitors from touching cattle and calves
- designate a specific visitor area to minimise contacts
- insist visitors wash their boots before entering and leaving
- supply rubber boots or plastic disposable boots and clean coveralls for visitors
- provide a footbath containing disinfectant
- insist workers wash their hands before milking cows and after working with sick animals
- insist workers wear protective plastic or rubber gloves for calving cows
- control the movement of dogs and cats between farms
- minimise the contact of dogs and cats with animals and feeding areas
- vaccinate farm dogs and cats for rabies and diseases common in your area
- wash farm clothing with detergents and bleach or washing soda.

Control Vehicles and Traffic Patterns on the Farm

Vehicles spread contaminated material on their tires, fenders and undercarriages. To decrease the spread of contaminants by vehicles:

- provide a separate laneway for use by the milk truck in accordance with the regulations of the Milk Act. This laneway must not be contaminated with manure

- provide "clean" routes for feed delivery vehicles-routes that are not contaminated by manure
- provide cattle with laneways that do not cross lanes used by milk trucks and feed delivery vehicles
- avoid sharing manure handling equipment with neighbours
- wash equipment to be shared with neighbours and insist on clean equipment coming onto your farm.

The most common means of contaminating feed or feeding areas is by on-farm equipment used for handling manure.

To decrease this risk:

- avoid using manure handling equipment for handling feeds, and if necessary wash before using for handling feed
- lay out feed storage and manure handling sites to avoid common traffic routes
- design and build barns where cows do not cross feeding alleys.

Control Feed and Feeding Equipment

Consider contaminated feeds (forages, pasture, grains and concentrates, water and waste milk), feeding equipment and systems when developing an on-farm biosecurity plan. The section on managing vehicles and farm traffic provides some basic information. The biosecurity of feeding should include plans to:

- purchase from suppliers with quality assurance and monitoring programs
- protect feeds from contamination through proper storage of chemicals, pesticides and medications
- protect feed from manure contamination
- establish storage facilities for feeds for various classes of livestock and systems to avoid errors in feeding practices
- harvest feeds at proper moisture contents and ensile them in suitable storage systems
- monitor water quality and assure clean delivery systems.

Clean Equipment

Disease can spread from animal to animal and farm to farm indirectly by small and large equipment. To reduce this method of spread:

- keep visiting vehicles out of areas accessible to livestock
- thoroughly wash and disinfect the inside, outside and tires of equipment shared with neighbours

- use a new disposable needle for each animal when administering treatments
- disinfect dehorners, hoof knives and trimmers after using on each animal
- use your own halters and clippers rather than borrowing them
- use separate shovels and forks for feeding and manure handling
- sanitize nursing bottles and buckets after each calf feeding
- maintain clean water troughs, water bowls, and feed mangers
- clean and sanitize equipment and materials used for handling deadstock.

Management of Groups and Housing

Young animals acquire infectious diseases through exposure with older infected or carrier animals. Housing and management systems, especially for dairy cattle, are constructed to minimise contact between young and older animals. In effect, the young are given time to develop immunity to diseases before joining the adults. The facilities also permit implementation of feeding and management practices to assure maximum growth, health and comfort. Dairy cattle owners implementing these strategies should:

- implement maternity-pen and newborn-calf management practices that prevent calves from ingesting manure.
- separate pre-weaned dairy calves from all other age groups
- house each milk-fed dairy calf in an individual pen or hutch
- place hutches away from dairy barn exhaust fans
- alternatively, house milk-fed calves in groups of less than eight calves
- house 4-8-month-old dairy calves in groups separately from older heifers
- house yearling and breeding age dairy heifers separately
- separate dry dairy cows from milking cows
- implement practices to prevent the spread of contagious mastitis (e.g. segregate cows with mastitis to the end of the milking order, sanitize milking equipment after milking a mastitic cow)
- separate replacement beef heifers from the cows
- move beef cows to a clean pasture, away from the wintering area, for calving
- organise chore routine to feed and milk isolated cattle after the main herd

- provide adequate pen, stall or bedded area per animal
- provide adequate feed bunk length and water trough access per animal.

Sanitation and Disinfection Management

Spread of disease is reduced when premises are clean and sanitary. In some cases, provincial legislation assures that minimum standards will be maintained. For example, the Milk Act (1987) regulates sanitation on dairy farms in Ontario. Several common management procedures assure adequate sanitation of farm premises.

Disposal of Dead Animals

Carcasses can be a hazard to people and other animals. They can contaminate soil, air and water and require special handling. To minimise property contamination and risk of spreading disease, owners should:

- dispose of dead animals within 48 hours of their death
- include all contaminated bedding, milk, manure or feed
- prevent scavenging by dogs, cats, birds, foxes, coyotes, wolves or bears
- wear protective clothing
- clean and disinfect the area that was occupied by the dead cattle.

Producers may choose on-farm disposal methods or have the dead animal picked up by a licensed collector.

On-farm disposal options under the Disposal of Dead Farm Animals Regulation [O. Reg 106/09] include:

- burial
- composting
- incineration
- disposal vessels
- anaerobic digestion (in some cases).

For burial, incineration or composting the Regulation describes minimum separation distances from a number of features which include:

- livestock housing facilities
- field drainage tiles
- residential and commercial lands
- surface water

- bedrock and aquifers
- wells (including municipal wells and floodplains).

The regulation also includes specific guidelines for each disposal option and limitations on the volume of deadstock disposed of for each option.

Producers also may transport their own deadstock to:

- common bins
- waste disposal sites approved under the Environmental Protection Act
- disposal facilities licensed under the Food Safety and Quality Act, 2001 (FSQA) [Regulation 105/09]
- a licensed veterinarian for post mortem.

Deadstock may be picked up by a collector licensed under the FSQA. Federal regulations apply to the transportation of dead cattle. Cattle carcasses or Specified Risk Materials (SRM) removed from a carcass must be dyed with a visible stripe. SRM are tissues that have been shown in infected cattle to contain concentrated levels of the bovine spongiform encephalopathy (BSE) agent. Collectors are responsible for applying the dye prior to removal of carcasses.

Producers transporting their deadstock or SRM to a receiver must follow guidelines in the FSQA regulations. To move dead cattle or SRM off-farm, producers must obtain a free 90-day permit from the Canadian Food Inspection Agency (CFIA). The process may involve an on-farm inspection by CFIA staff.

Cattle that die during transportation are considered to be SRM. The transporter requires a permit to bring the carcass back to the premises of origin or to a facility that is permitted by the CFIA to store, process or dispose of SRM.

CFIA has a comprehensive set of documents that outline the policy related to components that may not be added to feed (the enhanced feed ban).

Manage Manure and Control Flies

Infected animals often shed infectious agents in their feces, urine and other bodily fluids. The agents may contaminate feed, water and housing. To reduce the risk of spreading disease by manure:

- plan and install manure systems to prevent environmental contamination and comply with the Minimum Distance Separation II formula

- compost or store manure under conditions that destroy most disease-producing bacteria
- remove manure frequently from barns, yards and holding areas to prevent completion of life cycles by parasites and flies
- control the fly population by removing manure, using traps, baits or flypaper, using insecticides, using biological predators (wasps) or combinations of control measures
- store manure so it is inaccessible to cattle, especially young stock
- protect young stock from exposure to manure piles
- assure clean teats and udders for nursing calves by using clean calving areas and bedding packs with clean straw or shavings.

Manage Maternity, Dick and Calf Pens

Exposure of freshening cattle and calves to infectious agents is reduced by carefully managing maternity and sick pens. For disease control:

- use maternity pens only for freshening cattle
- use sick pens only for sick cattle
- prevent animal-to-animal contact between sick pens and maternity pens
- clean all manure from the pens after use
- disinfect the walls and floors after use
- allow the pens to dry after disinfection
- bed the pens well before the next animal enters
- if applicable, move cows and heifers to clean, dry pastures or paddocks for calving.

Use Disinfectants

Information about disinfectants is available on the product label or from farm supply dealers, veterinarians, the Canadian Animal Health Institute and the product manufacturers.

The Canadian Compendium of Veterinary Products, 2009, 11th Edition, contains the monographs of many common disinfectants. The indications for use, special properties, advantages, cautions and directions are described for each product. Your veterinarian should have a copy of this book in his/her veterinary clinic.

Familiarize yourself with the product information contained on the product label or package insert before making a selection. For a particular application, determine if a product:

- has activity against bacteria, fungi or viruses
- is active in organic debris (manure)
- is effective in hard water
- has decreased or enhanced activity in heat
- has residual activity for a period of time after application
- is compatible with soaps
- is caustic or has irritating fumes
- can be used on feeding equipment
- can be disposed of in accordance with provincial regulations and
- is appropriate for the intended use.

Several disinfectants for stables, housing and footbaths for visitors are shown in Table. These were obtained from the Compendium of Veterinary Products and are listed as examples, not endorsement. Other products may be available. Use the product information brochure included with the product to determine if the disinfectant meets the criteria for your application.

Disinfectants fall into six major categories: chlorhexidine, formaldehyde/glutaraldehyde, iodine complex, isopropanol, phenolic, or quarternary ammonium disinfectants. Several disinfectants fall into the other category not included in those mentioned above.

Foreign Animal Disease

For more than 50 years, Canada has successfully used border and import restrictions to prevent the entry of foreign animal disease (FAD). Ontario's livestock producers support these actions and supplement them with some common sense on-farm strategies.

Border Control

Canada prevents the introduction of FAD by strict border controls. The Canadian Food Inspection Agency and Customs Canada continue to:

- assess the risk of specific products that other countries want to import
- assess the animal health system in other countries
- ban the importation of ruminant (and some other) animals and their products
- ban the importation of used farm equipment
- use "sniffer" beagles to find food products carried by passengers

- seize and destroy foodstuffs and other illegal imports
- question passengers and inspect their baggage
- use disinfectants for passengers' shoes
- conduct specific tests and inspections of animals entering Canada.

On-Farm Control

Ontario's livestock producers prevent the introduction of FAD by common sense and practical farm-gate strategies.

For example, food and mouth disease (FMD) virus is easily killed by common procedures for cleaning or washing clothes-dry cleaning, bleach or washing soda. Experiments carried out 30 years ago showed that people examining the head area of clinically affected pigs harboured the FMD virus in their nasal cavity for less than two days. In these trials, infection of FMD was transmitted by snorting and coughing into the noses of steers within 30 minutes after examining the affected pigs. Presumably, the concept of a "stand-down period" after exposure to FMD virus came from these experiments. This confirms that persons who have been working with FMD animals must stay away from healthy animals for more than two days.

To prevent the introduction of foreign animal diseases from infected animals on farms in countries with diseases, Ontario's producers should:

- ask foreign visitors about diseases in their country of origin
- ask foreign visitors about their attendance on farms in their country of origin
- ask about and check the cleanliness of shoes and clothes of visitors
- provide rubber boots and protective clothing
- provide plastic boots
- restrict visitors from their farm if the visitor has been on a farm with a contagious foreign animal disease within the previous five days.

In case of an outbreak of a foreign animal disease in Ontario, federal veterinarians may impose bans on cattle movements to prevent the spread of contagious diseases from animal to animal. There would also be other control actions to stop the disease from spreading.

Disinfectants for Boot Wash

- Phenolic disinfectants (Beaucoup and Multi Phenolic Disinfectant) are examples of this class of disinfectant. Follow the label directions for mixing with water.

- Iodophors. Iosan is an example. Generally, prepare the boot wash by mixing 2-4 oz/gal of water (60-120 mL/4.5 L).
- Hypochlorites. Clorox or other brands of bleach that contain 5.25% sodium hypochloride. Mix 2-4 oz/gal (60-120 cc/4.5 L) of water.
- Chlorhexidine. Hibitane Disinfectant (2% w/v). Mix 120 mL per 3.8 L water.
- Sulfates. Virkon is an example. The common dilution rate for Virkon is 1% weight/volume. Mix a 50-gm packet in 5 L water.

Organic material (dirt and manure) inactivates many disinfectants. Therefore, clean your boots with a brush and water, removing all dirt and manure so that the disinfectant will sanitize the boots. Hypochlorites and iodophors will cause deterioration of rubber boots if left in contact. Prolong boot life by rinsing them with clear water after thoroughly disinfecting.

Conclusion

The work of disease prevention is never finished. Owners have the ultimate responsibility for herd protection. Visitors must respect biosecurity protocols put in place by livestock owners. Savvy livestock owners implement biosecurity strategies to prevent the introduction of disease to their herds and also to prevent the spread of diseases already present. To protect their herds, owners commonly:

- manage new arrivals
- manage and control farm traffic
- manage groups of animals and their housing
- sanitize and disinfect
- impose "stand-down periods" for visitors from foreign countries.

Review strategies for health protection and sanitation management of your herd using the lists above. Consult a veterinarian regarding which biosecurity strategies to use in your herd health program. Implement the appropriate strategies to insure health and comfort for your cattle. Make sure all workers and visitors are aware of their role in safeguarding the health of the herd.

For foreign animal diseases, border controls are a major part of the national biosecurity line of defence. On-farm biosecurity is an equally important line of defence. Together these steps minimise the entry and impact of diseases on your farm and in Canada.

Bioterrorism Involving Livestock

In light of the recent events in North America, governments and the public have a heightened awareness and concern about the possibility of bioterrorism involving agriculture, including livestock. This alert is not meant to cause panic or overstate the obvious. However, veterinarians and livestock owners may be the first to diagnose the early cases of a bioterrorist act in agriculture, as livestock can be sentinels of such an exposure. At the very least this note provides further points of contact for more information.

The Centres for Disease Control and Prevention in the United States has produced a list of the most likely biological agents to be used in an act of bioterrorism. Many of these are pathogens shared by humans and animals, and some have agricultural significance. Other agents of concern include Foreign Animal Diseases and other reportable diseases (such as Foot and Mouth Disease, Avian Influenza). Although many of the pathogens listed as possible bioterrorism agents would cause localised or sporadic disease, some may have serious public health or trade implications.

Veterinarians should remain current and vigilant for clinical signs consistent with any reportable or foreign animal disease, as well as unusual patterns or clusters of disease from the agents listed below. The Animal Health Laboratory (University of Guelph) and other diagnostic laboratories across Canada have been notified to be on alert for the detection of such agents. Diseases or clusters that might point to an exposure consistent with an act of bioterrorism could include: unusual demographic patterns, multiple species involvement, concurrent animal/public health problems, and unusual clusters of strains or serotypes.

Disinfection in on-Farm Biosecurity Procedures

Since the appearance of foot-and-mouth disease (FMD) in Europe in early 2001, its possible introduction into the United States has caused many livestock owners serious concern, so much so that many are now looking more closely at their biosecurity plans or their efforts to keep the disease out of their herds. Extension veterinarians have received many calls regarding which disinfectants to use on shoes, boots, tires, or other equipment in order to kill the foot-and-mouth disease virus. A few important points about disinfection should be made before choosing a disinfectant for routine farm use.

First, most disinfectants won't work if the surface to be disinfected isn't clean before applying the disinfectant.

Steam and high-pressure washers can be very useful to clean porous surfaces. Organic materials such as soil, plant debris (like straw), milk, blood, pus, and manure often inactivate some disinfectants or protect germs from the disinfectant's active ingredients. Chlorine-based disinfectants are especially subject to this problem. Chlorine, the active ingredient in bleach, is relatively quickly inactivated by organic debris such as manure, and even milk, at the concentrations usually used on clean surfaces.

In addition, even "hard" water can reduce or destroy the activity of some disinfectants. Likewise, some disinfectant solutions are only active for a few days after mixing or preparing. Failure to make a fresh solution of disinfectant after it has been prepared longer than a few days, or after it has become visibly contaminated by organic material like manure, may result in using a product that doesn't really work. Even worse, it may give a false sense of security. It is true that sufficient concentration and contact time can overcome some of these problems with certain classes of disinfectants, but often increasing the concentration or contact time makes use of the product impractical, costly, or caustic.

Disinfectants also vary considerably in their activity against the assorted germs bacteria, viruses, fungi, and protozoa about which livestock producers are concerned.

For example, plain vinegar (4% acetic acid) will readily kill the foot-and-mouth disease virus, but it won't do much to *Mycobacterium paratuberculosis*, the cause of Johne's disease. Most commonly used disinfectants are not active against bacterial spores, the environmentally hardy life form taken by the germs that cause tetanus, blackleg, botulism, and anthrax. Yes, formaldehyde is effective against most spores, but it is not really a practical disinfectant and is now considered a potential cancer-causing compound.

It is important to select a disinfectant that will be active across a wide spectrum of germs under the conditions in which it will usually be used.

These conditions include hard water, contamination with organic material, and potential for toxicity or damage to environmental surfaces or skin and clothing. It is also important to keep solutions clean and freshly made as directed by the manufacturer.

Lastly, disinfectants must have sufficient contact time with the surfaces to which they are applied in order to allow them to kill the germs with which we are concerned.

Contact time needed varies with the product and the germ. A quick splash of a dirty boot in a foot bath is not likely to accomplish anything except to give a false sense of security.

The chart of disinfectants on the next page has recently been made available by the USDA for field use in a foot-and-mouth disease outbreak ** and is useful in examining some of the previously stated points.

As you can see from the above, common household bleach would be an effective disinfectant for the FMD virus, but the recommended concentration (3% sodium hypochlorite) is 60% of full strength as it comes from the bottle. This concentration would damage clothing, shoes, and rubber goods and is mildly corrosive to steel surfaces. It can be used on an infected premise for FMD, but probably wouldn't be a good choice as a general purpose disinfectant for equipment and foot baths. Vinegar will also kill the virus, but wouldn't be a good choice for general use because of its lack of effectiveness against many other important germs. Obviously, lye is too caustic for general use.

On most farms, disinfectants will be used in foot baths or for cleaning equipment and livestock premises. The most commonly used disinfectants fall into the following classes.

Quaternary ammonium. The older quaternary ammonium compounds (Roccal D[T]) are good for some situations and relatively clean surfaces. They will not be particularly effective against FMD or *M. paratuberculosis*, the cause of Johne's disease, and have markedly reduced activity in the presence of organic material. Some of the newer quaternary ammonium preparations have improved activity. Compounds in this class usually have some detergent action; however, they are usually inactivated in contact with many soaps or soap residues.

Phenol-based compounds. These compounds are coal-tar derivatives and often have a strong pine-tar odour. They usually turn milky when added to water and have good activity in hard water and in the presence of some organic material. They are considered active against many bacteria, viruses, and fungi, including the bacteria that cause tuberculosis and Johne's disease. They are not especially active against the FMD virus; however, they are good all-purpose disinfectants for farm use. Some examples of this class of disinfectants include One Stroke Environ[R], Osyl[R], and Amphyl[R].

Hypochlorites. Chlorine compounds are good disinfectants on clean surfaces and have a broad spectrum of activity. They generally are more active in warm water. They can be somewhat irritating and can be harmful to clothing, rubber goods, and some metals. Some of the newer chlorine-based disinfectants are complex molecules that are less irritating and more effective than older ones such as bleach and Halazone[R]. Chlorine-based disinfectants are generally compatible with soaps but should never be mixed with acids. Their activity is strongly reduced by the presence of organic matter. Many chlorine solutions are unstable and need to be frequently replaced; read the label.

Iodophors. Iodine compounds have been used as antiseptics and disinfectants for many years. The iodophors are combinations of iodine and another molecule that makes them water-soluble. They are good disinfectants but are also not as effective in the presence of organic debris. Iodophors are generally less toxic than other disinfectants but can stain clothes and some surfaces. They are inactivated in the presence of some metals and by sunlight. They should not be mixed with quaternary ammonium disinfectants as they will be inactivated. Some examples of this class are Betadine[R] and Weladol[R].

Newer compounds. New disinfectants are being introduced regularly. Some of these are oxidising agents. Virkon S[R] is a peroxygen molecule/organic acid/surfactant combination (surfactants reduce surface tension to help water-based compounds penetrate). It appears to have a wide spectrum of activity against many kinds of germs (including the FMD virus). It is relatively stable in the presence of some organic material. It has a pH of around 2.6, when mixed as directed, but is labelled as nonirritating to skin. It is advertised as useful on many kinds of equipment, including saddles, brushes, buckets, etc. Another compound, based on peroxyacetic acid, is Oxy-Sept 333[R]. It is now EPA-approved for foot-and-mouth disease virus and is reportedly active against a broad spectrum of germs.

Remember, disinfectants are not to be applied to animals directly, unless labelled for such use, and you should consult the label to make sure there are no warnings against using them around feeders and in animal quarters. A general recommendation is to rinse disinfectants off after the appropriate amount of contact time if animals will have contact with the disinfected surfaces. Label directions should be strictly followed, and different classes of disinfectants should not be mixed.

In the event of a foreign animal disease outbreak, such as FMD, the type of disinfectant and procedures used in the cleanup of infected farms and for routine prevention activities will be selected by regulatory officials. For routine use in biosecurity programs at the farm level, producers should consider the major risks they are concerned about, consider the type of surface they wish to disinfect, the conditions under which the disinfectant will be used, and then select a disinfectant that best suits their needs. Information about activity in hard water or in the presence of organic debris, contact time needed, what germs are reliably killed, human use and environmental concerns, and other details are usually on the label or can be obtained from the company. Web sites are often good sources of information about individual products. Above all, producers should remember that disinfection is just one aspect of their biosecurity program.

Infectious Laryngotracheitis

A clinical case of Infectious Laryngotracheitis (ILT) has been confirmed in a poultry flock in Oxford County. The case was diagnosed on December 11 based on samples submitted to the Animal Health Laboratory at the University of Guelph. OMAFRA is working with the industry and the producer's veterinarian to reduce the risk of ILT spreading to other farms. Producers are instructed to heighten biosecurity. The service industries, as well as provincially and federally licensed processing plants, are also advised to enhance their biosecurity procedures. ILT is an acute respiratory disease caused by a herpes virus that can lead to devastating losses in the broiler and layer industries. The disease is most frequently associated with chickens, but can also cause disease in related birds such as pheasants and peafowl. Ontario has experienced ILT outbreaks in the past. The ILT virus does not cause infection in humans.

The mortality rate in affected flocks is usually low, although it can sometimes be more than 20%. Persistent shedding from recovered birds can prolong infection in a flock for a long period of time. The disease is normally characterised by gasping, neck extension, watery eyes (conjunctivitis), swollen sinuses and persistent nasal discharge. In severe cases, violent coughing may be observed, with blood evident in the trachea. ILT spreads relatively slowly through a flock within 2-4 weeks. The disease is usually spread by people, vehicles or equipment contaminated from an infected farm. Producers must be particularly careful with respect to the disposal of dead birds and manure, as the virus can be spread by these routes. Infected farms are advised to partially compost manure in the barn for at least 3 days prior to removal for complete composting on premises. Biosecurity is the most important means of prevention. The improper use of vaccines can result in clinical infection and further spread of the disease. Producers should consult their veterinarian to determine the appropriate biosecurity procedures for their operation.

Avian Influenza

Avian influenza (AI) is a contagious viral infection that can affect several species of food producing birds as well as pet birds and wild birds. AI viruses can be classified into two categories: low pathogenic (LPAI) and highly pathogenic (HPAI) forms, based on the severity of the illness caused in birds.

Avian influenza viruses, such as the highly pathogenic H5N1 virus present in Asia, may, on rare occasions, cause disease in humans. Transmission to humans has occurred through close contact with infected birds or heavily contaminated environments. In Canada, highly pathogenic avian influenza is a reportable disease under the *Health of Animals Act*, and all cases must be reported to the Canadian Food Inspection Agency (CFIA).

Avian Influenza (AI)-What to Expect if Your Animals are Infected?

About Avian Influenza

AI is a viral disease of birds. There are many different strains of the AI virus, and most of these have little or no affect on bird health. However, two types—known as H5 and H7—can cause severe illness and death in affected flocks. The Canadian Food Inspection Agency (CFIA) initiates disease response actions when birds are suspected to be infected with H5 or H7 AI viruses or other AI viruses determined to be highly pathogenic.

AI viruses can be classified into two broad categories: low pathogenic (LPAI) and highly pathogenic (HPAI), based on the severity of the illness caused in chickens. HPAI causes the greatest number of deaths in birds.

In Canada, Highly Pathogenic AI (HPAI) is a "federally reportable disease". This means that producers, veterinarians and laboratories must notify the CFIA of all suspected or confirmed cases. Birds affected by AI can show a variety of symptoms, including:

- high mortality and sudden death
- decreased food consumption
- huddling, depression, closed eyes
- respiratory signs (coughing and sneezing)
- decreased egg production
- watery greenish diarrhea
- excessive thirst
- swollen wattles and combs.

Humans are rarely affected by AI, except in a limited number of cases when individuals were in close contact with infected birds.

Nevertheless, public health authorities will take precautionary measures as warranted.

AI Disease Control

The CFIA takes immediate disease control actions in response to all situations where domestic birds are suspected of being infected with H5/H7 AI. While all disease response situations are different, the steps involved in an AI response normally include the following:

- movement restrictions, e.g., quarantines
- sample submission
- investigation
- destruction and disposal
- cleaning and disinfection
- compensation.

Quarantine

If you are a producer of a flock of birds that is suspected of being infected with H5/H7 or highly pathogenic AI, CFIA staff members will come to your farm to begin an investigation.

At that stage, a quarantine (also referred to as a Declaration of Infected Place) will be placed on your farm to control the potential spread of disease. You will be provided with documentation outlining the rules of the quarantine. Signs indicating that the property is a disease control area may also be provided. These signs must be posted at all entrances.

Under the quarantine, no birds, bird products (such as eggs) or bird by-products (such as manure) will be allowed to enter or leave your property without CFIA permission.

It is your responsibility to ensure people entering and exiting your property follow strict biosecurity measures, such as the cleaning and disinfection of footwear and vehicle tires. Other items being moved off the property will also need to be cleaned and disinfected under the direction of CFIA personnel.

While your property is under quarantine, you will have the following responsibilities:

- Request CFIA permission to move birds, bird products, bird by-products and poultry related equipment on or off the property.
- Ensure that footwear and vehicles leaving the property have been cleaned and disinfected.

- Apply strict biosecurity measures for yourself and any employees.
- Maintain fences and gates around the farm to control the movement of people on and off the premises.
- Where applicable, maintain fences and gates to areas housing birds—all birds should be housed indoors while the quarantine is in place.
- Report all sick and dying birds, and any that escape the farm.
- Implement vermin, feral animal, or wildlife control measures if warranted.
- Keep dogs, cats and other household pets confined.
- Inform all persons entering the farm of the quarantine.
- Limit on-farm visitors to essential services.

Investigation

The CFIA veterinarian will start the investigation by asking a series of questions about the health of your birds and the management practices you use. They will also attempt to answer any questions you may have. To help CFIA staff in their investigation, you may be asked to provide the following information:

- flock records (e.g., mortality, production, feed and water intake)
- veterinary records and laboratory reports
- a detailed description of farm management practices
- records of purchase/sale of feed, poultry, etc.
- movement on and off the premises during the past 21 days, (e.g. feed trucks, power, gas)
- farm visitor logbooks
- a site map of the farm
- contact information for the farm veterinarian.

Any premises considered to be in significant contact with your farm will also be quarantined because of the potential for disease spread. An example of a significant contact farm would be if you shared farm equipment or recently exchanged birds with another bird owner. Your cooperation and that of any other parties involved is critical to the success of the investigation.

Confirmation of AI

Testing to confirm if your birds are infected will be done as quickly as possible. Samples will be taken from the flock and sent to the nearest

CFIA-approved laboratory. Testing to confirm the 'H' type can be completed within one or two days.

If H5 or H7 or highly pathogenic AI is confirmed, all birds on your property will be humanely destroyed. Poultry products, such as eggs, will be destroyed as they are considered to pose a risk of spreading the AI virus to other birds. Flocks on other farms in the surrounding area will also be quarantined and tested.

If the particular virus present on your commercial farm is determined to be highly pathogenic, all the poultry on commercial operations within one kilometre of your commercial farm will also be pre-emptively destroyed. If H5 or H7 is confirmed, local health authorities will be available to answer any human health concerns or questions that you may have.

Destruction and Disposal

All birds are humanely destroyed using internationally recognised methods. Carcasses are disposed of in accordance with provincial requirements. Common disposal methods include burial, composting or incineration. On-farm burial must comply with provincial and municipal waste management requirements. All destruction and disposal costs may be covered by the CFIA.

Compensation

Under the Health of Animals Act, the CFIA may compensate owners for animals and things ordered destroyed during disease response situations. Compensation awards are based on market value, up to the maximum amounts established by the regulations.

Cleaning and Disinfection

Farms on which infected birds lived must be cleaned and disinfected after all destruction and disposal activities have been completed. The CFIA conducts a site assessment with the owner to determine which areas require cleaning and disinfection. The cleaning and disinfection process includes:

- removing litter and manure
- wet cleaning and disinfecting hard surfaces and structures
- cleaning and disinfecting tools and equipment.

Cleaning and disinfecting costs are your responsibility.

Removal of Quarantine

Once cleaning and disinfection are complete on the premises, the CFIA evaluates the farm to determine when the quarantine may be removed. Normally, this occurs 21 days after the cleaning and disinfection final inspection and approval.

Following removal of the Declaration of Infected Place, you can introduce new birds to the farm in accordance with CFIA requirements.

Confidentiality

As directed by the Privacy Act and other federal statues, the CFIA is required to protect private information collected. Any information provided by you during a disease response situation is treated as confidential, unless otherwise indicated.

Questions or Concerns?

If you have any questions or concerns related to the CFIA's disease response activities, please contact:

How to Prevent and Detect Disease in Backyard Flocks and Pet Birds?

Diseases such as highly pathogenic avian influenza and velogenic Newcastle disease can cause serious illness and death in many bird species. Fortunately, you can protect your birds and keep them healthy.

Follow five basic rules in the day-to-day care of your birds to reduce the risks posed by harmful diseases.

Prevent Contact with Wild Birds and Other Animals

Wild birds and other animals such as mice can carry a range of disease-causing viruses, parasites and bacteria. Make sure that your birds and their food and water are kept away from wild animals. Promptly clean up spilled feed and litter, and keep feed in sealed containers to avoid attracting unwanted guests.

Clean, Clean and Clean

Viruses, parasites and bacteria can live in organic matter such as litter and soil. Eliminate the risk of disease spread by routinely and thoroughly cleaning barns, cages, egg trays, gardening tools, and water and feed containers. No equipment should be shared with or borrowed from other bird owners. Always clean your hands, clothing and footwear before and after handling birds. Promptly dispose of dead birds, litter and unused eggs.

Spot the Signs and Report Early

Bird owners are legally responsible to notify authorities of serious bird diseases such as avian influenza. Call a veterinarian or a local office of the Canadian Food Inspection Agency if you suspect your birds are sick.

Signs to look for include:

- lack of energy, movement or appetite;
- decreased egg production;
- swelling around the head, neck and eyes;
- coughing, gasping for air or sneezing;
- nervous signs, tremors or lack of coordination;
- diarrhea; or
- sudden death.

It is always better to be overcautious. Report any bird that you think may be sick. Early reporting can greatly limit the effect of a disease on the health of your birds.

Limit Exposure to Visitors

People can spread bird diseases, too. As a general rule, do not give visitors access to your birds If someone must enter your property or handle your birds, make sure that their clothing, hands and footwear are clean and free of debris. Provide shoe or boot covers, or use a foot bath to prevent disease from entering or leaving your property. As well, the tires and wheel wells of any vehicles that have been around birds should be cleaned before entering your property.

Keep New Birds Separate when Entering Your Flock

Avoid introducing disease to your birds. New birds should be segregated and monitored for at least 30 days before entering your existing flock. Make sure that new birds come from reputable suppliers that have strict disease controls in place. Birds returning from shows or exhibits should also be segregated for at least two weeks

The Facts about Bird Flu

Avian Influenza (also known as "Bird Flu") continues to make headlines, sometimes causing confusion and anxiety among the general public. Is it safe to travel to Europe or Asia? Is it safe to eat poultry? Is it safe to go near wild birds?

Bird Flu, in its current form, is almost exclusively a disease of birds and does not pose a general threat to public health.

The reason that avian influenza and pandemic influenza have been linked together by some people is the potential for a virus usually found in animals to mutate and form a new virus that could easily infect humans, both directly and from one human to another. There is currently no human influenza pandemic in the world..

The following points may also help put Bird Flu into perspective.

- There are many different strains of Avian Influenza virus, commonly found in wild birds. Most cause no harm to birds or humans. The particular strain causing illness overseas is known as H5N1 (Asia).
- H5N1 (Asia) has not been detected in North America.
- Currently, H5N1 (Asia) does not cross easily from birds to humans. Almost all human cases to date resulted from direct contact with infected domestic birds or their droppings.
- There is no reason for the public to avoid eating properly cooked poultry and eggs.
- There is no evidence of cats infecting humans. A few cases of H5N1 (Asia) have been reported in domestic cats in Europe and Asia, but these were also caused by direct exposure to infected birds.
- The virus is likely being spread both by migrating wild birds and by the sometimes illegal movement of infected live poultry and on contaminated clothing and equipment. Wild birds may introduce the virus to a region or country, but poor biosecurity allows it to infect commercial poultry operations and spread to other farms.
- Because of migratory patterns, there is a chance that the virus will show up in wild birds this year in North America. The presence of the virus, however, does not necessarily mean that it will become widespread or that it will affect commercial poultry or humans. It certainly does not signal the start of a pandemic.

Notwithstanding all these facts, prevention and preparedness are appropriate at this time. The key to managing Bird Flu is controlling the disease in domestic poultry before it can become something more serious to humans as well as an economic disaster for the poultry industry. Another important step is minimising the contact, and possible mixing, among influenza viruses of birds, humans and other animals. Because of this, controlling the annual human flu virus through vaccination and good hygiene practices will also contribute to pandemic prevention, in addition to being a good public health measure.

What can be done to prevent the spread of Avian Influenza in Poultry?

- Governments are working with wildlife experts to detect the presence of Avian Influenza viruses in wild birds. Plans are in place for dealing with H5N1 if it is detected in wild or domestic birds in Canada.
- The commercial poultry industry is making biosecurity practices mandatory for its producers. It is important for all producers to follow these protocols routinely.
- There is currently no need to restrict or cancel poultry shows. However, bird owners should always observe proper biosecurity practices when attending any such event where birds from different sources are mixed.
- Bird owners, including hobbyists and those raising non-quota poultry, should follow these Biosecurity Basics to protect their birds from possible infection.

The 5 Biosecurity Basics for Bird Owners

- *Restrict visitor access.* People can bring diseases onto your farm, especially if they own or have been in contact with other birds. Contact with your birds should be restricted to those caring for them. It is a good idea to have separate clothing and footwear for use when dealing with your birds.
- *Prevent contact with wild birds and other animals.* Keep your birds in an enclosed or screened in area, and protect their food and water from contamination.
- *Don't bring disease home.* New additions or birds that have been at a fair or market should be isolated and observed for signs of disease for at least two weeks before joining the flock. Equipment such as cages should not be shared with other bird owners. Any shared equipment should be thoroughly washed and disinfected.
- *Keep it clean.* Routinely wash and disinfect cages, feed and water surfaces, boots and any equipment that comes in contact with the birds or their droppings. Wash your hands thoroughly after dealing with the birds.
- *Recognise and report illness.* Early detection is critical in successfully dealing with a disease outbreak. It is better to be overcautious than too late. If your birds show signs of disease, such as depression, abnormal egg production or feed consumption, respiratory problems, diarrhea, or sudden death, call your local veterinarian or contact the Canadian Food Inspection Agency immediately.

Infectious Bursal Disease of Chickens (Gumboro Disease)

This syndrome was first recognised near Gumboro, Delaware in 1962. It was first recognised in this province in 1970. It appears to be almost universal in broiler-growing areas.

Cause

Gumboro Disease is caused by a small, hardy virus, which shares many characteristics of the reovirus group. The organism is resistant to a great range of temperature and pH, but is killed by most disinfectants (formalin, cresol, iodine, etc.). The virus is rapidly spread by direct contact between birds and can survive for extended periods of time on inanimate objects, contaminated feed, etc. It does not appear to spread through the air.

Clinical Signs

Whitish, watery or mucoid diarrhea may be evident in the flock, with very sticky litter and soiling of vent feathers. Many birds may be reluctant to move with a tendency to sit. There is listlessness, dehydration and some deaths, with poor feed conversions. Secondary disease conditions, such as E. Coli infection, Marek's disease, gangrenous dermatitis and inclusion body hepatitis may increase in incidence, and condemnation rates may be elevated. The mortality pattern may range from the normal acceptable levels to a total of 15%, but the usual rate is low.

Four days after the onset of clinical signs, the mortality peaks and returns to normal within a week. The number of affected birds in a flock (morbidity) is variable and can approach 100%. Sick birds do not die if management is good and stresses kept to a minimum. Apparently subclinical disease can occur and destroy the birds immune system without causing obvious illness in a flock until secondary diseases develop.

Disease

The bursa of Fabricius, which is the organ responsible for disease protection in young birds, is the main target for the virus. Normally, the bursa of Fabricius regresses by early maturity. Hence, infectious bursal disease is most important in birds up to 4 months of age, and most critical between 2 and 4 weeks of age. The virus destroys many of the cells in the bursa of Fabricius and the resultant inflammation causes the organ to enlarge in size. Fever may develop and may cause depression and dehydration. Kidney damage (nephrosis) may result from the dehydration.

Hemorrhagic lesions throughout the body on the breast and leg muscles are seen on occasion. The post-mortem findings include pale, swollen kidneys, abnormally large or small bursa, hemorrhagic muscle lesions and possibly damage to the spleen, thymus or cecal tonsils.

Diagnosis

Diagnosis is made on the flock history and postmortem examination, and confirmed by virus isolation and identification. Serology and fluorescent antibody techniques are now available and help identify the disease agent. Histopathology of the bursa can also lead to a diagnosis.

Treatment and Control

No known chemotherapeutic or antibiotic agent is effective in the treatment or control of infectious bursal disease. Drug therapy is often inadvisable in the presence of severe kidney dmage. Electrolyte and/or multiple vitamin administration may be helpful in flocks where the disease is of relatively long standing and appetites poor. Good ventilation, warm temperatures and fresh water will help to reduce mortality. If secondary diseases become a problem, antibiotic therapy may be required, but this should be kept to a minimum.

After marketing a diseased flock, the farm should be completely depopulated of all species of birds. All litter and unused feed must be discarded and the building and equipment thoroughly cleaned and disinfected. Fumigation with formaldehyde is recommended if possible. (This is a hazardous procedure and must not be administered by inexperienced personnel.)

The building should be left vacant for 3 weeks. Vaccines are available in some countries, although they have not been introduced into Canada. Control of rodents, insects and wild birds is also important in the control of infectious disease.

Use of Livestock Medicines on the Dairy Farm

The use of livestock medicines on the dairy farm by producers and veterinarians is important for disease preventionand control. Management practices which prevent disease will reduce the need for drug treatment. Planned animal health and production programs commit the milk producer, the veterinarian and other herd advisors to the implementation of herd management policies which optimize health and production. However, when needed, medicines must be used responsibly.

This Factsheet provides recommendations on dairy cattle treatment, product label interpretation, treatment recording and antibiotic residue testing of milk. Information about the storage and handling of livestock medicines can be found in the Factsheet "Storage and Handling of Livestock Medicines on the Dairy Farm", Agdex 410/662.

Labelling

It is essential to read and understand drug labels to use medications safely and effectively. All livestock medicines bear labels which describe product information, indications for use, dosage, route of administration, warnings and storage instructions.

Product information includes:

- the product name ("brand name"),
- the name of the manufacturer or distributor,
- the drug identification number (D.I.N.),
- the manufacturers lot number,
- the active ingredient in the product, and
- the concentration of active ingredient.

Information provided by the label on product usage includes:

- the indications for medicine use, such as:
- the species (cattle, horses, swine)
- the class of livestock (lactating cows, nonlactating cows, calves) and,
- the disease conditions (mastitis, foot rot, metritis),
- the directions for medicine use, such as
- the dosage (how much, how often and for how long), and,
- the method of administration
    - o intramuscular-into the muscle,
    - o subcutaneous-under the skin,
    - o oral-by mouth,
    - o intramammary-into the udder through the teat end,
    - o intrauterine-into the uterus,
- storage requirements such as refrigeration, and
- the expiry date, the date past which the product should not be used.

All licensed products carry labels with warnings, cautions or precautions about the use of the product. For example, when an antibiotic product is recommended for use in cattle the label may include a warning statement "Warning: Not for use in lactating animals". This means that the product is for use only in non-lactating cattle such as dry cows and heifers. Products licensed for use in lactating cattle will indicate a with-holding time for milk following the last treatment. Products recommended for use in cattle will have a pre-slaughter withholding time on the label. With-holding times for milk and meat will only be valid if the product is used according to label instructions.

Labels may also provide information about disease management or product safety under the "Caution(s)" section. For example, a product recommended for intramuscular administration under directions for usage, may carry a caution that it is "not to be given intravenously" if the product is not safely given by that route. For further interpretation of the cautions, precautions or warnings on a product label consult your veterinarian. Inserts included with many products provide additional information such as side effects which may occur following product use in some animals. The inserts may contain scientific and medical terms which require interpretation by a veterinarian. Carefully read and understand all insert information before using the product.

Veterinarians must meet the same labelling requirements when drugs are dispensed in non-original containers. Information must be provided which identifies the product, the species and class of animals on which the product is to be used, directions for use and with-holding times. As well, the label must carry the name of the veterinary clinic and the veterinarian prescribing the product.

Drug labels are only useful to those who read them. Make a habit of reading the label before every use of a livestock medicine.

Extra-label Drug Use

"Extra-label" drug use is the use of a drug in any manner other than that listed on the label. Examples of extra-label use include:

- using a product to treat a lactating cow when the label does not list lactating dairy cattle
- using a product at a higher dosage than that listed on the label,
- using a product labelled for intramuscular injection as a subcutaneous injection, and
- using a drug to treat mastitis when the label recommends it only for use in the treatment of respiratory disease.

Extra-label drug use is permitted only under the supervision of a veterinarian. The with-holding time given on the medicine label does not apply when a drug is used in an extra-label manner. The veterinarian advising extra-label use of a livestock medicine is responsible for recommending a proper with-holding time for milk and meat.

Chapter 6

The Effect of Natural Toxins on Reproduction in Livestock

Reproductive efficiency is the most important economic factor in livestock production. Thus, the hypothalamo-pituitary-gonadal regulatory axis, accessory sexual organ functionality, and the complex events involved in fertilization, implantation, and embryonic and fetal development may be sensitive to therapeutic agents, environmental pollutants, and natural toxicants. There are many factors that adversely affect reproduction, one of which is toxic substances in the diets of animals. Toxic materials can affect reproductive success by causing abortions, interfering with libido, estrus, oogenesis, or spermatogenesis, causing emaciation and subsequent abnormal mating behaviour, birth defects, and increasing the time between parturition and rebreeding.

Examples of natural toxicants in poisonous plants interfering with reproduction are numerous. Abortion in livestock from locoweeds, ponderosa pine needles, broom snakeweeds, fescue, and others are reported in studies. Selenium and seleniferous forage inhibit estrus in cattle and swine. Emaciation and temporary illness from sneezeweeds, bitterweed, locoweed, larkspur, lupines, and others may interfere with mating. Embryonic loss and birth defects from Veratrum, lupines, locoweeds, poison hemlock, and so on, may occur. As suggested, toxins have many diverse and economically adverse effects on reproductive performance in livestock.

Colostrum for the Dairy Calf

The cost of raising replacement dairy animals increases if calf-rearing results in higher-than-normal mortality or requires medicine to treat preventable diseases. At birth, a calf has a poorly developed immune system. The placenta does not allow the transfer of antibodies, also known as immunoglobulins (Ig), from the mother to the fetus during pregnancy. True colostrum is the "first milk", which is rich with the antibodies that provide the calf protection from diseases in early life until the calf's own immune system becomes functional. Colostrum is also important as the first source of nutrients after birth.

Antibodies are proteins that identify and destroy disease-causing organisms, or pathogens, in the calf. Three major types of Ig (G, M and A) are typically found in the colostrum of dairy cows in percentages of 85%-90%, 5%-10% and 5%-10%, respectively.

The three types of immunoglobulins have specific roles in the immune system. The primary role of IgG is to identify and help destroy invading pathogens. IgG can move out of the bloodstream and into other areas of the body where it helps identify pathogens. The principal role of IgM is to identify and destroy bacteria that have entered the blood. IgA attaches to the membranes that line many organs, such as the intestine, and prevents pathogens from attaching and causing disease.

Timing

Timing is critical to a successful colostrum-feeding management program. The ability of a calf's small intestine to absorb immunoglobulins drops rapidly over the first few hours of life. By 24 hr of age, the ability to absorb immunoglobulins is nearly nonexistent. If a calf has not received any colostrum within 12 hr of birth, it is unlikely to be able to absorb enough antibodies to have adequate immunity. For this reason, a calf should receive the first feeding of colostrum within 1 hr of birth when possible.

Passive immunity is the temporary protection that the calf receives from the cow through the transfer of maternal antibodies present in colostrum. Passive immunity protects the calf until its own immune system becomes active. With active immunity, an older calf is mature enough to produce antibodies in response to vaccinations or to fight infections that it is exposed to.

The goal for proper colostrum feeding is for the calf to achieve a minimum blood serum IgG level greater than 10 mg/mL. As an alternative to testing for serum IgG levels, serum total protein levels may be measured. A serum total protein level of 5.2 g/dL is considered equivalent to the serum IgG level of 10 mg/mL.

"Failure of passive transfer" (FPT) occurs when the acceptable levels of total protein or IgG are not achieved by 24-48 hr after birth.

Canadian and American reports often show that as many as 35%-40% of dairy calves suffer from FTP, which indicates that many calves have inadequate immunity and are more likely to get sick. A recent study reported serum total protein levels in Ontario dairy calves; approximately 35% of calves had FPT.

It is accepted that the higher the concentration of IgG or serum total protein levels in the calf 48 hr after birth, the greater the protection the calf has against disease pathogens. A study by the U.S. National Animal Health Monitoring System found that calves with low immunoglobulin levels in their blood 2 days after birth had a death rate over the next 8 weeks that was more than twice that of calves with acceptable levels of serum immunoglobulins.

Quantity

The best practice is to feed 4 L of high-quality colostrum to Holstein calves within 1 hr of birth. A second feeding of 2?3 L of colostrum should be given within the next 8 hr. Calves that are bottle-fed colostrum have a better chance of receiving enough immunoglobulins than calves left to nurse from their mother. Calves that fail to drink on their own within 3 hr of birth should be given colostrum by esophageal feeder.

Quality

The immunoglobulin content in the first "milking" colostrum is typically 5%-6% (50-60 g/L), but can range from less than 2% (20 g/L) to greater than 15% (150 g/L). The concentration of antibodies in colostrum decreases rapidly with each milking as the transition from colostrum to milk production occurs. Usually, the second milking has 65% as much immunoglobulins as the first milking. By the third milking, the level has fallen to 40% of the first milking. Some differences in the composition between true colostrum, transitional milk and whole milk are shown in Table.

Nutrients in colostrum, such as fat and protein, are also important to the calf for growth and development. The lactose concentration in colostrum is less than is present in whole milk, which reduces the chance of diarrhea in the newborn calf.

Dairy breeds produce lower concentrations of immunoglobulins in their colostrum than beef breeds. Among the common dairy breeds, Holsteins tend to have the lowest level and Jerseys the highest.

First-lactation heifers usually have lower levels than older cows that are in their third or greater lactation. Cows that have had less than a 4-week dry period usually have lower levels of antibodies.

Cleanliness

A significant challenge of colostrum feeding is keeping it clean. While feeding colostrum is essential to provide passive immunity for the calf, it is also one of the first ways to potentially expose the calf to such pathogens as *E. coli*, *Salmonella* or *Mycobacterium avium paratuberculosis*, the bacterial species responsible for Johne's disease. Pathogens can also cause diseases such as scours and septicemia and may interfere with passive absorption of the antibodies from the gut into the circulation system. Clean udders, milking equipment and calf-feeding equipment well before harvesting, storing and feeding colostrum.

Feed calves colostrum that has a total bacteria count of less than 100,000 colony-forming units (CFU)/mL and a total coliform count of less than 10,000 CFU/mL.

An Ontario study in 2002 found that 12% of colostrum samples had high levels of bacteria. The study concluded that the source of the bacteria was from either a dirty udder or dirty collection container. Similarly, results from a recent U.S. study reported bacterial results from the colostrum on 12 dairy farms. In that study, the average total plate count was 16.1 million CFU/mL, and the total coliform count was 2.7 million CFU/mL, indicating that dairy farmers should pay more attention to colostrum cleanliness.

Storage

Colostrum can be refrigerated cold at 1°C-2°C (33°F-35°F) for up to a week or kept frozen at -20°C (-4°F) for up to year. Avoid frost-free freezers. Two-litre plastic containers or freezer bags are ideal. If using freezer bags, double-bag the colostrum.

Thaw the colostrum in 50°C (120°F) water. Do not thaw at room temperature, as bacteria double every 20-30 min. at room temperature.

Colostrum can also be thawed in a microwave, with care. Microwave the colostrum on low for short periods of time. Avoid hot spots and mix partially thawed containers if necessary.

Recent research at the University of Minnesota found that colostrum can be heated to 60°C (140°F) without damaging the antibodies. However, when the colostrum was heated to 63°C (145°F), the antibodies were reduced by 34%.

Do not mix colostrum from two or more cows.

Colostrum can also be pasteurised on farm to reduce the presence of pathogens. The newer pasteurisation equipment is convenient to use. It can provide a very effective method of reducing pathogens to improve colostrum cleanliness and dramatically reduce the potential for diseases to be transmitted by feeding colostrum. It is important to keep pasteurisation equipment clean and to follow manufacturers?

directions for both temperature and time of treatment so as not to destroy the antibodies present in the colostrum. Pasteurising colostrum significantly reduces the level of bacteria present. However, to reduce bacterial regrowth, pasteurised colostrum must be properly stored, as described above, until it is fed. As a final alternative, if sufficient amounts of high-quality colostrum are not available, commercial colostrum supplements can be a valuable tool for increasing calf immunity or for use as a disease-management strategy when transmission of disease from the mother to the calf is an issue.

In the U.S., some colostrum supplements use bovine serum as the source of immunoglobulins. These products have not been approved in Canada, and their ability to achieve satisfactory immune status in dairy calves is debatable. Maternal colostrum replacement products prepared from colostrum from healthy cows are available in Canada, and published research supports its effectiveness at increasing immunity protection in dairy calves.

Follow manufacturer's directions with these products.

Testing

Testing equipment can be purchased from a number of farm suppliers. A colostrum hydrometre, often called a Colostrometre®, estimates the immunoglobulins in the colostrum by measuring the specific gravity of colostrum. Accurate measurements require that the temperature of the colostrum be correct, typically, at room temperature. Higher or lower temperatures result in incorrect readings. Follow manufacturer's directions. Some manufacturers provide information to adjust readings if tested at other temperatures.

Readings of more than 50 g/L for colostrum IgG concentrations are desirable.

Refractometres are commonly used by veterinarians to evaluate the degree of passive transfer in calves. Refractometres are used to estimate the total serum protein level in the calf. Readings of 5.2 g/dL or greater indicate successful passive transfer.

Summary

Colostrum is the critical first step to the health and survival of newborn calves. The successful transfer of the antibody protection in colostrum from cow to calf is based on four key factors:

- How quickly the calf receives colostrum after birth - within 1 hr is best.
- How much colostrum the calf receives - 4 L at first feeding.
- The immunoglobulin concentration in the colostrum.

Sanitation for Fly and Disease Management at Confined Livestock Facilities

Flies

The stable fly and house fly are the major insect pests at confined livestock units. The stable fly has a piercing-type mouthpart which is used to pierce the skin to obtain a blood meal. House flies do not bite because they have a sponging-type mouthpart with which they feed on semi-liquid material. The life cycles of the two species are similar, consisting of eggs, larvae (maggots), pupae, and the adult. During summer months the stable fly completes its life cycle in about three weeks and the house fly in about two weeks. Both species deposit eggs in wet, decaying organic matter. This includes spilled livestock feed and manure mixed with soil and moisture. In addition, the house fly will breed in fresh manure. The two species generally overwinter as slowly developing larvae in breeding areas below the frost line. The house fly may breed during the winter in warm buildings if breeding material is present.

Cattle under attack by stable flies will bunch together with each animal attempting to find a position within the bunch which protects their front legs—the favoured feeding site of the flies. Considerable energy is expended by foot stomping, tail switching, and throwing the head down toward the front legs in an effort to dislodge the flies or prevent feeding.

Stable flies reduce weight gains, milk production, and feed efficiency—both from their feeding and because of the bunching behaviour of cattle which may induce or increase heat stress.

House flies have not been shown to reduce animal weight gain and feed efficiency but are known to transmit several animal diseases. The disease organisms recovered from house flies range from viruses to nematodes. The most common of these are the bacteria associated with enteric infections. The house fly mouthparts and feeding habits (filth sources) make it efficient in transmitting bacterial and viral agents. Over 100 different disease organisms have been recovered from house flies, and the fly has been implicated in the transmission of 65 of these. Transmission may simply involve the mechanical transfer of the disease agent from the mouthparts or body of the fly to the animal host. In other cases, the disease agent may multiply in the fly and be transmitted after populations of the disease agent build up to high numbers or it changes to a different life stage.

Stable flies, because of their blood-feeding habits, have also been suspected of transmitting diseases, but most research in disease transmission studies with the stable fly have proven negative. Mastitis in dairy herds is one of the exceptions. The mastitis organisms are routinely spread by the stable fly to the teat ends of heifers or lactating cows. Mastitis, the most costly disease of dairy cattle, is caused when the bacteria invade the teat and gain entrance to the mammary gland.

One other important economic factor associated with stable flies and house flies is the threat of nuisance lawsuits. Generally odour, dust, and flies are cited together as constituting a nuisance by the plaintiffs. The lawsuit may seek damages or, perhaps worse, request closing of the livestock facility.

Effective house fly and stable fly control cannot be achieved with insecticides alone. Proper animal manure management and sanitation must be the major element in a good fly control program. Confined livestock facilities should be designed or modified to facilitate ease in cleaning and to minimise accumulations of manure, spilled feed and other sources of organic debris.

Breeding Areas

In feedlots, major fly breeding areas include: 1) fence lines where manure mixed with wet soil accumulates; 2) along feeding aprons where, because of the slope of the aprons, moisture and manure accumulate; 3) at the edges of potholes, in pen corners, and around gates; 4) along pen drainage channels or edges of holding ponds at the water-soil interface unless it is sloped enough to drain and dry quickly; 5) along the sloping edges of mounds where moisture runoff occurs; 6) around waterers if leakage or spillage occurs; 7) at the edge of and under feed bunks; 8) in and around feed handling facilities if the feed becomes wet; 9) at the edges of stored manure, if the edges are loosely packed and wet; and 10) at the edge of silage and haylage.

In addition, excessive moisture may provide fly-breeding areas under round bales in and around old hay or straw stack butts and in sick pens and horse stables if hay or straw bedding is used.

Fly breeding areas at dairies might include any of the areas described for feedlots, but fly breeding might also occur under and around self-feeding forage racks and in calf hutches or pens where hay or straw bedding is used.

Fly breeding at swine units is restricted primarily to the house fly. Stable flies will breed in wet, spilled feed around swine units but are rarely, if ever, found breeding in wet soil-swine manure mixtures as is the house fly. The most common fly breeding areas will differ to some extent with the type of facility but basically occur at any location where swine manure or wet feed stuffs are allowed to remain for a period of 10-14 days. In swine confinement buildings, where manure is collected below the slatted floor pens, a crust occasionally occurs unless some type of agitation is provided.

House flies may breed in the crust just below the surface. Agitation provided by the manure dropping from the slatted floor will suffice if enough liquid is available and the depth of the drop is greater than one foot. Manure may also accumulate in the upper corners just below the slatted floors. The same situation is true for cattle confinement buildings.

In modified open front or open lots, the main fly breeding areas may be under and around the self-feeders or in the corners of the facility. While the accumulation of the manure or spilled feed may be relatively small, an area about a square yard in size will allow several thousand flies to develop in a period of two to three weeks.

Fly Management

Reduction of fly breeding areas in feedlots is dependent primarily on manure management and keeping the lots dry. Mounds are a key element in this process. They should be built and maintained to provide a dry area for the cattle and drainage for the excess moisture to move from the pens to the drainage system. During wet periods, the wet edges of the mounds can be scraped into the lot in a thin layer to facilitate rapid drying.

Maximum stocking rates create tramping action that helps in drying. The lots also can be dragged periodically which helps maintain a dry surface. The area behind the feeding apron should be scraped at two week intervals and the manure either removed or spread out in the lot for rapid drying. Drainage systems should be maintained with enough slope to move the moisture to the holding ponds rapidly which allows for rapid drying as well.

Haylage and silage piles may have drainage at the edges. The seepage provides an excellent fly-breeding site. Covering this seepage area with black plastic should create enough heat to kill the developing fly larvae. If manure is used as fertilizer and spread directly on farm fields, care should be taken to spread the manure thin enough for rapid drying. If the manure is spread at depths of three to four inches or more and enough moisture is present, it may allow fly breeding.

Water tanks should be surrounded by a concrete apron and equipped with a drain line to facilitate cleaning without creating a muddy area in the lot. Float valves on waterers should be protected to prevent animals from causing an overflow. Livestock pens usually have enough organic matter present to create a fly breeding area wherever water accumulates.

Feedlots designed or modified to meet the Environmental Protection Agency?s pollution runoff standards can have an additional fly breeding area in the debris basin. The purpose of the debris basin is to intercept the feedlot runoff, allow the solids to settle and channel the liquids into the holding pond. The basin should be sloped enough to prevent water from standing and provide quick drying. Solids should be removed regularly to prevent fly breeding.

If spray mists are used to cool cattle or hogs or to settle dust, care should be taken to prevent puddles from forming. Dragging the surface of the lot may fill in low areas where puddles form.

Each livestock unit is different and there may be fly breeding occurring in only two or three locations. However, since even small amounts of fly-breeding material can support large numbers of flies, these areas should be located and removed. Manure management and sanitation can be expensive but should be considered a required management practice in livestock production. The benefits (better animal performance, more efficient use of insecticides, better working conditions for employees, more attractive facility for commercial customers, reduced risk of nuisance lawsuits, and reduced chances of disease outbreaks) may offset the expense.

Accurate Detection of Bruising Improves Carcase Value & Welfare

Females and older animals are more susceptible to bruising at slaughter and more accurate detection methods may identify who is economically responsible for losses in carcase value due to bruising. There are a number of external causes of bruises that are sustained during the last hours and days before beef animals are slaughtered and animal factors, such as sex and age, may contribute to the development of bruises, at least in some cases. Just some of the findings of a review by Dutch and Chilean scientists, led by Wageningen University's Ana Strappini.

"Better understanding is still needed of the biological mechanisms accounting for the higher bruise rates in females and older animals," she said, adding that it was also clear that beef cattle sold through markets can suffer bruising that could have been avoided by transporting animals directly from the farm to the slaughterhouse.

Studies of bruises, as detected on carcases at the slaughterhouse, may provide useful information about the traumatic situations the animals endure during the pre-slaughter period.

Many aspects of cattle transport contribute to bruising. Transport conditions, such as stocking density and duration of the journey, seem to have more effect on bruising than distance travelled. "But finding an optimal stocking density for livestock transport under different conditions is still a contentious issue," said Dr Strappini.

"Bruised tissues may store historical information about the harmful situations that the animal underwent prior to slaughter. The farmer and the transport companies have economic incentives to prevent and reduce bruising. However, slaughterhouses do not have simple and accurate methods for post-mortem age estimation of bruises to assess accurately when bruises were sustained.

This is a relevant problem, due to the importance of having to decide who is economically accountable for the losses. Although the number of bruises, their anatomical location, severity and even the healing process might offer a rapid tool for identifying and evaluating the circumstances during the pre-slaughter period such as high stocking density, rough handling or inappropriate facility infrastructure, other sensitive techniques should be considered for refined assessments of the time the bruises were incurred."

"More investigation of the time between bruising and slaughter may help to clarify the risk factors that have contributed to the occurrence of bruises and will also help to identify the risks for animal welfare," she added.

The modern diagnostic techniques applied when evaluating human bruises may be studied for bovine bruises as well. "Immuno-histochemistry and cytochemistry seem to be promising methods to be applied to measure morphological or biochemical changes, which can clearly be distinguished from non-bruised tissues."

"But age assessment of bruises continues to be a crude process. A wide variety of factors intrinsic to the animal can influence the inflammatory process and subsequent repair.

"Normal biological variation among animals is therefore bound to result in substantial overlap among proposed time frames in the healing process."

Greenhouse Gases and Environmental Impacts of the Livestock Industry

We have a severe effect on grazing land and livestock production systems Pierre Gerber, Livestock Policy Officer Animal Production and Health Division at the FAO, told the recent World Meat Congress in Cape Town, South Africa.

He said it is a great challenge for food security and the security of the public's health and biodiversity and it is also a challenge for the livestock industry.

Dr Gerber said that at present 3.4 billion hectares around the world is devoted to pasture land - 26 per cent of the total of "emerged lands". While this was being used to a low intensity in the developing nations, the intensity of production was increasing in the Latin American countries.

A large part of this pasture land is too dry or too cold for crop use, and only sparsely inhabited.

While the grazing area is not increasing on a global scale, in tropical Latin America there is rapid expansion of pastures, which is encroaching into valuable ecosystems, with 0.3 to 0.4 per cent of forest lost to pasture annually, Dr Gerber said.

Ranching is a primary reason for this deforestation. About 20 per cent of the world's pastures and rangeland have been degraded to some extent - maybe as much as 73 per cent in dry areas.

This growing intensity of production is reflected in the fact that global production of meat is expected to more than double form 229 million tonnes in 1999/2001 to 465 million tonnes by 2050 and milk production in the same period will go up from 580 million tonnes to 1,043 million tonnes.

The bulk of this growth will be in the developing countries - Brazil, China, India and Russia in particular.

At the same time the global demand for livestock products is also expected to double.

The total area dedicated to feed crop production is 471 million hectares, equivalent to 33 per cent of the total arable land and most of this is in the developed countries.

Recent reports from the UN and other organisations show that land use changes that have followed deforestation contribute 18.3 per cent to total greenhouse gas emissions, while agriculture accounts for 13.5 per cent (six per cent of which is agricultural soils and 5.1 per cent is livestock and manure) and the transportation sector accounts for 13.5 per cent with 10 per cent down to road transport.

Dr Gerber said that estimates of greenhouse gas emissions for the livestock sector are substantial when the different forms of emissions throughout the livestock commodity chain are taken into account.

Emissions come from feed production, such as chemical fertilizer production, deforestation for pasture and feed crops, cultivation of feed crops, feed transport and soil organic matter losses in pastures and feed crops. The emissions also come from animal production, such as enteric fermentation and methane and nitrous oxide emissions from manure and as a result of the transportation of animal products.

Dr Gerber said that livestock production contributes about nine per cent of the total carbon dioxide emissions produced by human activity, but 37 per cent of methane and 65 per cent of nitrous oxide emissions.

In total there are 7.1 billion tonnes of CO_2 produced through feed and livestock production in the world and he said that CO_2 releases resulting from fossil fuel consumption used for the production of feed grains (tractors, fertilizer production, drying, milling and transporting) and feed oil crops also need to be attributed to livestock.

The same applies to the processing and transport of animal products.

Relative Greenhouse Gas Contributions Along the Food Chain

Dr Gerber told the conference that there is a similar impact from the livestock sector on water usage throughout the world.

The livestock sector accounts for about eight per cent of the total water used and most of this is used on feed crops and irrigation of pasture land.

There are also consequences for the water supply through the pollution of water by manure and pathogens and contaminants from livestock and freed production and there are consequences for the erosion of the soil through grazing animals and intensive rearing of animals.

The growing intensity of livestock production is changing the biodiversity of the world and is having an effect of different ecosystems.

"Climate change is a big problem, but there are other issues as well," Dr Gerber said.

He said the industry and governments need to control land use and control the carbon and nitrogen in cultivated soil. He said it is necessary to manage livestock production better for productivity gain and also for manure management.

He added that given the projected expansion of the livestock sector, major corrective measures need to be taken to address the environmental impact of livestock production, which will otherwise worsen dramatically.

Growing economies and populations, together with the increasing scarcity of environmental resources and rising environmental problems, are more and more demanding enhanced environmental services, such as clean air and water, and recreation areas.

He said that while there will be an increasing intensification in livestock production, the challenge will be to make the process environmentally acceptable.

With extensive land-based production the decision makers will have to ensure that it includes environmental parameters particularly in vulnerable areas.

Dr Gerber added that policy makers need to provide a framework for landscape maintenance, biodiversity protection, clean water and eventually carbon sequestration from extensive grazing systems, in addition to that for the production of conventional livestock commodities.

"Policy decisions have to take into account environmental and health aspects of livestock production," he said.

"And the FAO is endeavouring to raise awareness of these problems."

A strong political will and urgency, together with the identification of potential contributors and beneficiaries, are required to initiate action and investment, Dr Gerber concluded.

Plant-Produced Protein Eases Dairy Cow Infection

Coliform mastitis is the most prevalent form of clinical mastitis in the dairy industry and is mainly caused by the *Escherichia coli* bacterium. Use of antibiotics to control mastitis infections can be expensive and carries with it concerns about the emergence of antibiotic-resistant bacteria. Yet mastitis is expensive too, costing dairy farmers an estimated $2 billion annually from incapacitated cows and milk that can't be sold.

Now, ARS molecular biologist Lev Nemchinov and plant pathologist Rosemarie Hammond have endowed a potato virus with a gene that—when introduced into a host plant—prompts the plant to produce a therapeutic protein called "CD14." First isolated by Dante Zarlenga of the ARS Bovine Functional Genomics Laboratory, this beneficial protein can be extracted from the plant and used to treat mastitis.

Nemchinov and Hammond are with the Molecular Plant Pathology Laboratory at Beltsville, Maryland. Their engineered, or recombinant, virus is called "PVX/CD14," for "potato virus X carrying the gene of therapeutic protein CD14."

Protein Power

The researchers extracted enough CD14 protein from the inoculated plants for field tests. Their colleagues, dairy scientist Max Paape and microbiologist Douglas Bannerman with the ARS Bovine Functional Genomics Laboratory, Beltsville, Maryland, tested the purified protein's ability to alleviate mastitis in dairy cows.

CD14 is known to help the immune system fight infection, but it is present in the cow's mammary gland at low levels. CD14 binds to a molecule known as "lipopolysaccharide," located on *E. coli*'s outer membrane. The scientists hypothesized that increasing the level of CD14 in the milk would enhance protection.

"When inserted into the mammary gland through the opening of a cow's teats, CD14 binds to the *E. coli* and triggers the cow's immune response, which fights mastitis inflammation," says Hammond. "That process helps to neutralise and clear toxins produced by the bacteria, lessening the chances of an excessive immune response."

The researchers infused the CD14 protein into one of a test cow's four teats, or "quarters." All four quarters were subsequently exposed to *E. coli*. Fewer viable bacteria were recovered from the quarter that received the CD14 treatment than from those that did not receive the plant-derived protein.

A Top-Notch Protein Factory

The researchers chose tobacco plants, *Nicotiana benthamiana*, to be their CD14-producing factories. "We inoculated the young plants with laboratory-produced RNA of our recombinant virus by rubbing a small drop of the RNA onto the plants' leaves."

Once the viral RNA enters the test plants through small tears in the leaves, it begins to spread and the CD14 gene begins to make the protein. The protein can then be extracted from mashed-up leaves.

Hammond says this is the first report of a functionally active animal receptor protein being produced in a plant.

When a plant is infected with a virus, thousands of copies of viral RNA are made in each plant cell. "When our recombinant virus reproduces itself in a plant cell, it also makes the target protein, CD14. This is how the virus turns a plant into a bio-factory that rapidly generates proteins of interest," says Nemchinov. Separating and extracting the therapeutic protein from the host plant is possible because Nemchinov first tagged the CD14 protein with the amino acid histidine. "The histamine-tag tracker enables us to harvest high levels of the desired plant-produced protein," he says.

The researchers are able to purify about 1,000 micrograms of CD14 from 10 grams of leaf tissue taken from one plant. That means each of these plants provides enough protein to potentially treat about 10 cows with a dosage of 100 micrograms each. Fifty plants would yield purified protein to treat a herd of 500 cows.

One large greenhouse can accommodate enough host plants for large-scale purification of therapeutic protein. Another advantage of the greenhouse approach is that the plants inoculated with laboratory-derived infectious RNA could be used to inoculate more plants to scale-up protein production.

Partnering for Further Development

ARS has applied for patent protection on the plant-derived CD14, and the researchers are now seeking partners to help further develop and test the protein for safety, effectiveness, and proper dosage.

The CD14-based product may eventually be commercially developed for use by dairy farmers as a treatment to prevent cows from becoming infected during their dry period. Dairy cows are milked for 305 days and then enter a 60-day dry period, during which they are most susceptible to coliform infections. The plant-made CD14 could be incorporated into a polymer and infused into the udder during dry-off. The polymer would allow slow release throughout the dry period to help fight infections.

"As a next step, we'd like to work with a commercial partner to produce large volumes of plant-made CD14 and then conduct further tests to determine the most effective dosages in cows to achieve maximal protection from infection," says Hammond. "We may not need as much as the 100 micrograms per dose that we used in our first tests. Further studies may show that we can achieve the same results with less protein."

By expressing an easily purified therapeutic protein in plants, the ARS team has developed a novel preventive approach to treating mastitis that may provide a cost-effective alternative to antibiotic use.

Hyperspectral Imaging: A Non-Invasive Technique

Consumers have shown a willingness to pay a premium for guaranteed tender steaks. To increase consumer satisfaction and value of beef, the industry has a strong interest in tenderness predictors. An accurate, noninvasive, online tenderness instrument is needed for packing plant scenarios. Since beef carcasses are quality and yield graded by USDA employees two days postmortem, and product typically reaches the consumer at 14 days postmortem, the machine would need to accurately predict the ultimate 14 day postmortem tenderness value.

Hyperspectral imaging is a technique whereby multiple reflectance images are captured at regular intervals along a spectral axis. Thus, each pixel in a hyperspectral image has spectral reflectance data. In contrast, near-infrared (NIR) spectroscopy measures spectral reflectance of an entire field of view rather than a single pixel. Thus, hyperspectral imaging would be expected to be much more accurate as a result of the additional information that is captured.

The objective of this research project was to develop and validate an accurate, noninvasive tenderness predictor by scanning steaks at 14 days postmortem to then ultimately develop a system to predict the 14 day tenderness level (tender, intermediate, or tough) by scanning steaks at two day postmortem.

Procedure

Hyperspectral Imaging Apparatus

A hyperspectral imaging apparatus was constructed by integrating a CCD digital video camera (Model: IPX- 2M30, Imperx Inc., Boca Raton, FL) and a spectrograph (Model: Enhanced series Imspector, Specim, Finland). The spectrograph has a spectral range of 400-1,000 nm. Spatial and spectral calibrations were performed. A diffuse-flood lighting system was designed using tungsten-halogen lamps and a dome with a white reflectance coating. Lighting was provided with six 50-W tungsten halogen lamps (Model: MR16, Phillips Lighting Co.). A lamp controller (Model: TXC300-72/120, Mercron Industries, Richardson, Tex.) converted 60 Hz AC voltage to 60 kHz. At this high frequency, tungsten halogen lamps do not respond quickly. This simulates a constant DC voltage power supply. Over the lifespan, tungsten halogen lamps get dimmer. A photodiode was placed near a tungsten halogen lamp that provides feedback to the controller. Based on the feedback, the current input to tungsten halogen lamps is increased to provide a constant intensity output. Over the lamps, a hemispherical dome of 40 cm diametre was placed, providing uniform diffuse light over the steak.

USDA Choice and Select grade longissimus steaks from between the 12th and 13th ribs and cut to 1-inch thickness were placed on metal trays which were then vacuum packaged. The trays contained 6-14 steaks and were placed in a commercial refrigerator for a 24 hour thawing time to an internal temperature of 1-6°C. Steaks and a white reference plate were then placed on a Teflon-coated plate mounted on a linear slide that used a stepper motor for movement.

The steak was then scanned by the camera to obtain a three-dimensional data cube (reflectance by two dimensional position). Scanning takes approximately 30 seconds to collect the image, and each file is approximately 600 mb. Images were obtained at wavelength intervals of 2 nm. Steaks were then cooked immediately on an impingement oven to an internal temperature of 69.5-72.2°C, and slice shear force values were obtained within one minute by an Instron Texture Analyser.

Statistical Analysis

A 200 by 300 pixel region of the image was selected for analysis. The region of interest was in the approximate location where slice shear force samples were obtained. Principal component analysis was carried out to reduce the dimension along the spectral axis. Over 90% of the variance of all bands in the image was explained by the first five principal components. The first four principle components are shown. On each principal component image, co-occurrence matrix analysis was conducted to extract eight image-textural features; thus a total of 40 image-textural features were actually obtained from each steak. To reduce the number of features and predict 3 tenderness categories (tender - slice shear force 21 kg; intermediate 21.1 to 25.9 kg; tough 26 kg), a canonical discriminant model was developed. Leave-oneout cross validation procedures were implemented to predict the tenderness level.

Results

The model correctly classified 9 tender, nine intermediate, and five tough samples, incorrectly classified three intermediate samples as tender, and incorrectly classified one tender sample as intermediate. All tough samples were correctly identified. Tenderness was predicted by this hyperspectral imaging device with 96.4% accuracy (Table 1).

Implications

This hyperspectral imaging system was effective in accurately predicting 14 day tenderness of beef longissimus steaks. With implementation of a non-invasive, accurate tenderness predictor, beef cuts could be labelled and sold at a premium as guaranteed tender. With this premium, producers, feedlots and packing plants would reap the benefits together.

Pelvic Measurements and Calving Difficulty

Calving Difficulty

Calving difficulty results in major economic loss to the beef cattle industry. Estimated losses resulting from dystocia (calving difficulty) equal or exceed $750,000,000 annually. Calving difficulty influences the economics of a cow/calf enterprise through increased calf death loss, increased labour and veterinary costs, reduced subsequent reproductive performance of the cow, potential loss of the cow, and reduced milk production.

Calf mortality may be four to eight times greater in dystocia cases than in normal births. The majority of calf deaths occur within the first 24 hours following calving (58 percent), with 75 percent of the total occurring within the first week of life. Studies indicate that calf death loss due to dystocia accounts for the single largest perinatal and postnatal death loss category through the first 96 hours after birth. A number of factors affect calving difficulty, including:

- Birth weight of the calf
- Pelvic area of the cow
- Gestation length
- Sex of calf
- Inadequacies in heifer development
- Body condition of the cow at calving
- Abnormalities in hormone profiles at the time of birth
- Abnormal presentation of the calf at birth.

We also know that the single major cause of dystocia is a disproportion between size of the calf at birth (birth weight) and the cow's birth canal (pelvic area). Differences in pelvic area are generally due to pelvic height, with discrepancies between the dam and fetus more likely to occur for pelvic height and depth of calf chest than for width measures.

Pelvic size, independent of cow weight, affects calving difficulty. Heifers of increased skeletal size usually have larger pelvic openings, but also tend to have heavier calves at birth. Hence, selection for cow size alone is ineffective.

Heifer weight and age generally have a positive relationship to pelvic area, but weight is not always a good indicator. External dimensions such as width of hooks and length of rump are not good indicators of pelvic area or calving difficulty. For these reasons pelvic measurements can be a useful management tool to eliminate heifers with a higher potential for calving difficulty.

Pelvic Measurements

University of Nebraska researchers developed ratios that you may use to estimate deliverable calf size. You can divide total pelvic area prior to breeding by a ratio that is based on age and weight to estimate the amount of birth weight a heifer could accommodate as a 2-year-old without substantial difficulty.

Example: A 600-pound yearling heifer (Table 1) with a pelvic area of 140 sq cm should be able to deliver, as a 2-year-old, a 67-pound calf without difficulty (140/2.1 = 67).

Pelvic measurements can be obtained at the time of pregnancy exam, but a factor of 2.7 should be used to estimate calf birth weight of 18- to 19-month-old 800-pound heifers. Tables 1 and 2 provide estimates of the deliverable calf size a heifer can accommodate at first calving based on pelvic area at given weights and ages. Scientists at the University of Nebraska suggest that these ratios appear to be good indicators of dystocia and report an accuracy of nearly 80 percent.

Structural traits in cattle tend to be highly heritable and pelvic area is no exception. This means there is a large genetic influence on pelvic area, which results in rapid response to selection. However, pelvic area is genetically correlated with many other traits, so selection for increased pelvic area alone can result in other traits changing for the worse. For example, selecting for increased pelvic area can result in increased birth weight and mature weight.

Pelvic measurements can be taken prior to the first breeding season and combined with a reproductive tract examination. Pelvic measurements should be used in addition to, not in place of, selection for size, weight, and above all, fertility. Producers should be aware that selection for pelvic area is likely to result in increased size of the entire skeleton and animal. Increased skeletal size of the dam will be reflected in higher birth weight and dimensions of the calf.

Pelvic measurements, on the other hand, can be used to successfully identify abnormally small or abnormally shaped pelvises. These situations, if left unidentified, are often associated with extreme dystocia, resulting in Cesarean delivery and even death of the calf or cow.

Pelvic measurements can be obtained with a Rice Pelvimetre, manufactured by Lane Manufacturing, 2075 South Valencia, Unit C, Denver, CO 80231; the Krautmann-Litton Bovine Pelvic Metre, manufactured Jorgensen Labs, Inc., 2198 West 15th St., Loveland, CO 80538; or the Equibov Bovine Pelvimetre, manufactured by Equibov, 205 Harris Street, Rockwood, Ontario, CAN NOB2KO.

The vertical measurement is the vertical diametre between the symphysis pubis on the floor of the pelvis and the sacral vertebrae. The horizontal measurement is obtained by determining the horizontal diametre at its widest point between the left and right ileal shafts. These measurements are read in centimetres and multiplied together to obtain the total pelvic area in square centimetres.

Measurements may be obtained by a veterinarian or an experienced producer. It is important that the person doing the measuring have a thorough understanding of the birth canal, pelvic structure, and reproductive tract. Practice and experience are necessary before accurate measurements can be obtained.

Summary

Calving ease will continue to be an important consideration as the industry produces fast-growing muscular progeny by terminal sires. These sires should be selected on measures of direct calving ease by using EPD (expected progeny difference) values for calving ease and birth weight.

To accommodate fairly heavy birth weights, scientists at Colorado State University recommend that you develop a cow herd that excels in maternal calving ease. Sires of replacement females should be selected to maintain cow size and milk production at levels compatible with available resources.

In addition, cows should be selected for total maternal calving ease along with gestation length.

Research indicates that bigger is not necessarily better when one considers actual pelvic measurements. In other words, heifers with large pelvic measurements fail to calve more easily than average-sized heifers. However, heifers with abnormally small pelvises or abnormally shaped pelvises generally experience a higher than normal incidence of calving difficulty and should be identified and culled from the herd.

Remember, pelvic area and shape are only a part of the calving difficulty complex. Follow the suggestions in the following list to minimise the incidence and severity of calving difficulty in your herd. How to reduce calving difficulty (in ranked order):

1. Breed heifers to proven calving ease bulls (low birth-weight EPDs)
2. Develop heifers to prebreeding target weights.
3. Ensure that heifers are in good body condition going into the calving period (minimum body condition score of 5).
4. Obtain pelvic measurements at yearling age and cull heifers with abnormally shaped or abnormally small pelvic areas.

A Blueprint for Eradicating Bovine TB

Over the course of the year 2008, bovine tuberculosis infected over 500 new herds in the UK and led to the slaughter of over four and a half thousand cattle. Despite of continuous prevention strategies, this figure presents a sharp increase from the year before. The economic cost of the disease has also soared in recent years, from £7.3 million in the year 1998/9 to £32.6 million in 2007/08.

Struggling to get to grips with the ever-worsening effects of the disease, the TB Advisory Group published a final report on the possibility of complete eradication within the UK.

Released April 8 2009, the report - Bovine Tuberculosis in England: Towards Eradication - April 2009 - is a conclusion of almost three years work in which the Group says it played a key role in obtaining stakeholder buy-in to TB control policies, independently challenged Government and considered issues of concern to stakeholders whilst advising on practical implementation of control policies.

In the Direction of a Wind of Change

The UK recently saw the proposed legislation of a badger cull fail on the grounds of animal welfare. This has left many in the role of TB control feeling deflated and at a loss for ideas of where to turn next.

The report emphasises the need to reinstate a sense of urgency, ensuring sufficient resources are available for this to become possible. It also makes it clear that there is 'no magic bullet' for eventual control and eradication. All attempts must be made to minimise the disease transmission risk and a consistent risk reduction approach must be used for all breakdowns.

According to the report, a "holistic multifaceted approach" is needed that uses a combination of control measures. For the TB Advisory

Group's plan of action to succeed they say that substantial extra costs must be incurred and a realistic time frame between 10 and 20 years must be allowed for any hope of complete eradication.

The plan must also take into account TB existing in wildlife reservoirs. The report says that the plan will need to "stop the spread of Bovine TB from existing endemic areas", and also "Stamp out the disease where it occurs in new areas."

Another area that the reports highlights is to dispel the myths and misconceptions that many farmers have over the nature of the disease. "Who communicates this message is paramount and veterinary endorsement is key", says the report. Similarly the importance of this goal can not be lost due to set backs and promises of an easier future. A badger vaccine will not provide an instant cure, says the report. It will be used merely as part of the multifaceted approach.

Most importantly, clear leadership from both the government and the industry needs to materialise.

Chapter 7

Animal Productivity Diseases

In growing pigs, research has shown that negative and inconsistent handling increases fear responses.

One study by Prof Hemsworth looks at growth rates in pigs. His study showed that the growth rate of positively handled pigs was 455 g/ dy, whereas it was only 404 g/day in pigs negatively handled. The growth rates for inconsistent pigs was 420 g/day.

In this situation, the growth rate was reduced due to the animals stress response (cortisol concentrations were elevated in inconsistent and negatively handled pigs), said Prof Hemsworth.

A similar study was carried out in laying hens, looking at the effects of negative handling (sudden and unpredictable movements in front of the pens) and positive handling (an extra two minutes spent in front of the cages, and slow deliberate movements).

Hen time at the front of the cage was measured, with less time at the front of cage took as avoidance of human contact. Stress responses and egg production were also measured. The results show that positive handling of birds means that the hens were keen to have increased human contact (less fearful), spending more time at the front of the cage.

The corticosterone stress levels were much higher in hens handled negatively, than in positively handled hens.

Subsequently egg production in the hens was eight per cent higher in hens that had a positive human-animal relationship.

The number of studies across species with strong correlations between stress and negative handling, leaves no doubt that negative handling evokes stress, affecting animal welfare and production, said Prof Hemsworth.

Negative Handling: Animal Health

As mentioned above, studies have shown that negative handling affects an animals fear of humans, leading to stress, which consequently affects health.

One study shows that socialised (used to positive human contact, so less fear) birds had higher feed conversion efficiency. In this same study, birds that were less socialised (more fear of humans) had higher lesions and deaths, as well as overall poorer health. Prof Hemsworth says this is due to the response of immune system, suppressing antibodies when an animal is stressed, leading to poor health.

A different study, carried out on veal calves, looked at average daily gain and mortality. Fast movements by stockpersons were negatively associated with daily liveweight gain. As well as this, negative behaviour was seen to increase mortality, although unit size was a large variable in this study.

Longer flight distances in dairy cattle, also have a positive correlation with lameness in dairy cattle. In a study of 36 dairy heifers, 48 per cent of heifers with a flight distance of 4.8m (to humans) were lame, and on average milk yields were 1.3 kg/day less, than those with a shorter flight distance.

To compare, of cows with a flight distance of 2.81m (those that were less fearful), only six per cent showed signs of lameness.

On pasture based systems, Prof Hemsworth said that lameness in dairy cows was affected by two significant factors-the condition of farm tracks and farmers' patience in the dairy, ie. their behaviour towards cows.

What is an attitude, asks Prof Hemsworth. It is something that affects our behaviour and although they are stable and resistant, attitudes are learnt.

They are shaped through direct and indirect experiences, therefore throughout one's lifespan, attitudes can be changed.

Professor Hemsworth says to change stockperson behaviour, it is important to target attitudes as well.

Cognitive-behavioural Training

Evidence from studies carried out by Prof Hemsworth and colleagues suggests that stockperson training can improve animal productivity and welfare.

To change the behaviour of stockpeople towards farm animals ultimately requires:

- targeting the beliefs that underlie the behaviour,
- targeting the behaviour in question, and
- then maintaining these changed beliefs and behaviours.

One study looked at the benefits of cognitive-behavioural training on dairy units. The key variables measured were stockperson attitudes, stockperson behaviours, fear of humans and animal productivity.

The study demonstrated that training significantly improved stockperson attitudes. Training also halved stockperson negative behaviour towards cattle. Whilst the fear responses of cows didn't change much, the changes were small but significant, said Prof Hemsworth. Flight distance did not decrease a lot, but milk cortisol levels did often decline.

A second study looked at the effects of training on cow productivity at 94 dairy farms. After training, which targeted stockperson attitudes and behaviour, milk yield per cow increased, as did the protein and fat content of the milk produced.

Stockperson Selection

Some studies, by the Animal Welfare Science Centre, have looked at how the selection of a stockperson can improve animal welfare.

Measuring stockperson characteristics prior to employment, may give employers an idea of stockperson attitues and consequently behaviour towards animals.

Concluding, Prof Hemsworth said that the role of a stockperson in animal welfare and productivity should not be underestimated. The studies in this article are just a few of the many that have been carried out. The outcomes highlight how important the role of the stockperson is in developing positive human animal relationships, and consequently improving animal welfare and productivity.

Naturl Alternative to Antibiotics in Animal Feed

Binding molecules from seaweed extracts within the molecular layers of natural clays makes the clay ten times more efficient at adsorbing the harmful fungal toxins which are commonly found in animal feed. The resulting new hybrid product has a huge potential market worldwide as a completely natural, effective alternative to the formerly-used antibiotics, which are now prohibited from this use in the European Union.

EUREKA project E! 3025 Monlisa has developed a new, completely natural alternative to using antibiotics in animal feed. Formerly used as growth promoters and to prevent the damaging effects of the fungal toxins which are often present in feed, antibiotics have been prohibited for use in animal feed in the EU since 2006 – creating a need for a new, effective antimycotoxic agent. The already-known capacity of clay minerals to adsorb hese mycotoxins has been increased ten times over in this project by the incorporation of molecules of extracts from algae (seaweed) onto the clay. The resulting activated clay is added to animal feed, where it is highly efficient at adsorbing the mycotoxin molecles.

Mycotoxins – which have, over the years, become one of the major concerns of animal feed producers – are the chemical products of fungal moulds which are often present in cereal grains and forage crops. They are especially prevalent in warm and humid regions like Asia and South America, where the climate conditions favour fungal grwth. Mycotoxins cause a range of problems to farm animals, including reducing their feed intake and therefore reducing their growth. Consming mycotoxins can also make animals more susceptible to disease and lead to damaged liver and kidneys. Human health is affected by consumption of animal products like meat, milk and eggs which are contaminatedwith mycotoxins.

For many years, a range of antibiotics have been applied in low doses to animal feed, to control the productin of mycotoxins and to act as growth promoters. But as part of its policy to reduce the threat of developing microbial resistance to antibiotics, the use of antibiotics in animal feed was banned in the EU from 1 January 2006. This policy mean that there was an urgent need for some other way to controlthe mycotoxins in animal feed.

Multiple Applications

French company Olmix was aware that some naturally occurring clay minerals, particularly montmorillonite, were able to adsorb organic molecules including these toxic substances onto their surface. The clays are formed of layers of about one nanometre. Within the Monalisa project, French company Olmix employed the process known as intercalation, to include molecules of seaweed extracts between the clay layers; making the new product Amadeite®. This process separates the layers (delamination) and increases the interlayer space; and as a result the clay is able to adsorb up to ten times the amoun of mycotoxins compared to pure clay.

Dr Anca Laza-Knoerr says: "Our idea was to introduce a natural product into the clay, rather than the synthetic polymers that had already been used to make activated clays." The process, which involves suspension of the clay and seaweed extract in water, followed by removal of the remaining algal extract, centrifugation and drying, now holds two worldwide patents. It is the development of this process that is the main technological innovation. As well as applications in the field of animal feed, the clay-algal extract combination has potential uses in the purification of water from waste effluents, heavy metals or radioactive pollution.

Olmix worked on the preparation of the seaweed for extraction in partnership with other specialised laboratories: the European Research Centre for Algae (CEVA) and the Laboratoire de Matériaux à Porosité Contrôlée (LMPC, Mulhouse). The seaweed is readily and cheaply available on beaches but it has to be carefully washed before extraction so that residual sand does not damage the extraction machinery. Use of the algae for this purpose has the secondary benefit that it is so plentiful in some regions that it is considered an environmental pollutant.

Successful Products from Basic Research

Further development of the application of Amadeite® in animal feed was made by the Spanish partner in the Monalisa project, Adiveter, which has specialist knowledge of the microbiological monitoring of feeds and raw materials. As a result, since the end of the project Olmix has been able to market two products for use in animal feed. The first, called M Feed, is a growth promoter which aids the digestive process and helps to maintain the balance of the gut microflora. The economic yield of the animals is increased by enhanced digestion and growth. The second product, MTx+ adsorbs harful mycotoxins from the feed and its use has been shown to increase livestock productivity. Both products are already being marketed in many countries throughout Europe and in Russia, the US and Japan, with estimated worldwide sales of €12 million in 2009.

Aurélie Garel, also from Olmix, said: "The activated clay Amadeite is an innovative, natural additive, and really important for agriculture and many other industrial applications."

Developing the new activated clay product Amadeite was only possible through the support of EUREKA, which helped to locate and bring together partner organisations with the right expertise and facilities. Collaboration between the partner organisations was excellent. The impact of the project is that it allowed the development of a highly effective range of products, to take the place of the antibiotics whose widespread use was threatening to increase microbial resistance.

Economic Impacts of Foreign Animal Disease

This report by the USDA Economic Research Service presents a modelling framework in which epidemiological model results are integrated with an economic model of the U.S. agricultural sector to enable estimation of the economic impacts of outbreaks of foreign-source livestock diseases. It was produced by Philip L. Paarlberg, Ann Hillberg Seitzinger, John G. Lee and Kenneth H. Mathews, Jr.

Summary

This study uses a modelling framework to estimate the nature of economic impacts of outbreaks of foreign-source livestock diseases. The model is more comprehensive than previous work because (1) it has components for modelling both economic effects and disease-spread effects from an outbreak, for which the results can be integrated; (2) it assesses the effects of a disease outbreak on major agricultural sectors—livestock and crops—along vertical market chains, from production to consumption; and (3) it projects the impacts of the disease outbreak over 20 calendar quarters, rather than for just 1 year.

What is the Issue?

As more is learned about the impacts of foreign animal disease outbreaks, questions arise regarding the efficiency of existing animal disease-impact models for capturing the array of effects across many economic sectors and over time. Previous models lacked adequate treatment of either the economic components or the epidemiological components, and, in some cases, both. Further, there is a need to address the ways that alternative control strategies affect the economic interests of the numerous agricultural sectors, both on and off the farm.

What are the Major Findings?

While the framework can be applied to many livestock diseases, this study demonstrates the model with a hypothetical outbreak of foot-and-mouth disease (FMD). The outbreak is assumed to occur in small hog operations in the U.S. Midwest, a result of using contaminated garbage as feed. Various disease-control strategies are entered into the model. The economic effects from each control strategy are based on 50 iterations of the disease-outbreak epidemiological model, which randomly assigns different herd sizes, spatial distribution, and other variables for each iteration. The model produced the following key results:

- Epidemiological model results show that relatively few animals need to be destroyed because of the disease.

- Economic model results show large monetary losses for beef, beef cattle, hogs, and pork sectors, mainly caused by the loss of exports under a given set of foreign sanitary and phytosanitary policies. Other agricultural sectors experience small losses or, in some cases, small gains.
- Swine and pork sectors recover shortly after export restrictions end, while effects on beef and cattle sectors last longer due to the longer cattle production cycle.
- Disease control strategies that reduce the duration of the outbreak are the most effective choices for reducing the economic toll. The model found that three strategies reduce the duration to less than one quarter. In order of least to most effective, based on the mean number of days to end the outbreak, for the hypothetical outbreak these are:
  - o Destruction of only those herds within a radius of 1 km that have had direct contact with infected herds: Outbreaks average 56.48 days.
  - o Direct-and-indirect-contact slaughter, which destroys direct contact herds plus those herds indirectly exposed to an infected herd through movement of people, vehicles, or other possible sources of infection: Outbreaks average 54.99 days.
  - o Destruction of all herds within a 1 km radius of the initial outbreak: Outbreaks average 36.8 days.
- Export embargoes increase domestic meat supplies, and domestic consumers benefit from lower prices during the quarters in which exports are embargoed.
- Model results, extended to 16 quarters, show that for this hypothetical outbreak:
  - o After 7 quarters, production of all commodities increases to the point where both domestic consumption and trade return to predisease levels.
  - o Total trade losses, plus other disease-related costs to capital and management, amount to between $2,773 million and $4,062 million, compared with a disease-free baseline period (2001-2004).

How was the Study Conducted?

The framework has two components: (1) The North American Animal Disease-spread Model (NAADSM), developed by the U.S. Department of Agriculture's Animal and Plant Health Inspection

Service, which enables estimates of epidemiological damages (supply shocks) from varying diseasespread and control scenarios. These estimates can then be integrated with (2) an economic model, developed by Paarlberg, Seitzinger, and Lee, that assesses effects of supply shocks from the epidemiological model, along with demand and trade shocks, projected over the simulation period.

To illustrate the modelling framework, a hypothetical outbreak of FMD arising from feeding garbage in four small farrow-to-finish operations is examined under three alternative control strategies and three levels of disease intensity. The control strategies are (1) destruction of direct-contact herds, (2) destruction of direct-contact and indirect-contact herds, and (3) destruction of all animals within a 1-km ring. Disease intensity is examined at low, medium, and high levels. Each disease control scenario was simulated 50 times, with the epidemiological model determining effects of an outbreak.

Animal losses and duration of the FMD outbreak are sensitive to the conditions assumed for the outbreak, i.e., that it started on small pig farms and was confined to them. Alternative scenarios could result in higher costs than reported in this analysis. Future work will evaluate the robustness of these results.

Variation in Costs for Animal Disease Prevention

A new report *Cost of National Prevention Systems for Animal Diseases and Zoonoses in Developing and Transition Countries* commissioned by the World Organisation for Animal Health (OIE) has been published recently. The main conclusions are summarised by Jackie Linden for TheCattleSite. These include that there is a strong relationship between current spending per animal unit and GDP, although this is unrelated to the level of spending required.

The OIE commissioned Civic Consulting to conduct a study on the cost of National Prevention Systems (NPS) for animal diseases and zoonoses in developing and transition countries.

The final report explains that the aims of the study were two-fold:

1. to estimate the 'peace time' costs of Veterinary Services allowing early detection and rapid response to emerging and re-emerging diseases in different regions, economies, animal health systems and eco-systems, and
2. to develop economic indicators within the OIE-PVS Tool for the Evaluation of Performance of Veterinary Services (OIE-PVS Tool).

The study was based on the results of in-depth research in nine OIE member countries (Costa Rica, Kyrgyzstan, Mongolia, Morocco, Romania, Turkey, Uganda, Uruguay, Viet Nam) and an extensive analysis of possible economic indicators.

The study team, led by Dr Frank Alleweldt (project director) and Professor Martin Upton (lead author of the economic analysis), drew 11 main conclusions.

NPS Expenditure is Related to Livestock Population

One leading conclusion in the report is that substantial differences in the public expenditure for the National Prevention System for Animal Diseases and Zoonoses exist between case study countries, reaching from 10 million international dollars to 167 million international dollars. The average expenditure on the National Prevention System was 48.6 million international dollars in the baseline year, 2007.

Variations in expenditures between case study countries are clearly associated with differences in livestock population. Operational costs of the National Prevention System, when expressed on a per Veterinary Livestock Units (VLU) basis, therefore give a comparative measure of the level of service provision in relation to the quantitative requirements.

NPS Expenditure is Correlated with GDP

In the case study countries, there is a close relationship between Gross Domestic Product (GDP) and the total public expenditures for the National Prevention System. Differences in GDP explain to a large degree the variation in NPS expenditures.

NPS expenditure appears to be mainly dependent on the country's ability to pay, rather than on the veterinary requirements. This may lead to a significant under-funding of the NPS, most notably in low-income countries. In these cases, veterinary services require a higher priority in the national budget allocation, and/or sustained external support to be able to effectively address global animal health challenges.

Strong Relationship between Spending per Animal Unit and Incomes

Differences in NPS expenditures between countries on a per VLU basis are, at least partly, explained by differences in per capita incomes, according to the report.

While the overall average NPS cost per VLU for the seven countries amounts to 5.66 international dollars, the average for the three low-income countries – Uganda, Kyrgyzstan and Viet Nam – is only 3.82

international dollars. The average for the two lower-middle-income countries – Mongolia and Morocco – is 5.28 international dollars, while that for the upper-middle-income countries – Costa Rica and Turkey – is 8.79 international dollars.

Total Cost Per Animal Unit is not Affected by the Level of Centralisation

Sub-national expenditures tend to increase relative to the centralised expenditures with increasing size of the national territory. Operating expenditures associated with the National Prevention System are incurred either centrally, in or near the main centre of government, or dispersed more widely in provincial, regional or district locations. A high central expenditure in Costa Rica is clearly associated with a centralised structure in a relatively small country, whereas Turkey, Morocco and Vietnam, three of the largest countries in area, spent about three-quarters of the total NPS operating expenditure at the sub-national level. Provided that both central and regional elements are included, the average total cost per VLU may be unaffected by the extent of decentralised expenditure, say the report's authors.

No Explanation Found for Differences in Spending Patterns

Spending patterns for different categories of expenditures varied between the case study countries, however, this provides little explanation for differences in overall NPS expenditures. Levels of staff costs and expenditures such as travel costs appear to be directly related to levels of per capita income of case study countries.

Considerable differences in spending that depend on other factors are related to three categories:

- fees for private veterinarians conducting public service mission (up to 0.96 international dollar/VLU)
- expenditures for vaccines (up to 1.57 international dollar/VLU), and
- compensation of livestock holders (up to 0.74 international dollar/VLU).

In some other countries, spending for these items is zero or close to zero.

Public NPS Expenditure is not Related to the Strength of the Private Sector

There is no evidence that a stronger private veterinary sector reduces public NPS expenditures in the case study countries. The relative strength of the private veterinary sector, expressed as the ratio of public to private veterinarians, appears to be related to the income level of the country.

In the case study countries, both NPS expenditures and the relative importance of the private veterinary sector increase with a higher GNI per caita.

GDP Does not Correlate with NPS Requirement

The strong linear correlation between GDP and NPS expenditures for the case study countries can be used to estimate current National Prevention System expenditure. However, this approach provides a rough estimation of the likely current level of funding of the NPS only, and does not in any case determine the optimal level of NPS expenditures in a given country.

The only reliable and accurate method of obtaining data on NPS expenditures in other countries currently available is by means of direct measurement, using the methodology developed for this study.

System to Quantify Results is Needed

The report's authors suggest that a quantitative expression of OIE-PVS Evaluation (OIE-PVS Tool for the Evaluation of Performance of Veterinary Services) results would be helpful for assessing the degree of compliance with OIE International Standards on Quality of Veterinary Services in a systemic perspective.

In future refinements of the PVS Tool, the introduction of a more quantitative approach could be considered. Also, due to the cross-cutting character of several of the critical competencies used for the PVS Tool, it is currently difficult to correlate the costs for key NPS elements (e.g. veterinary diagnostic laboratories) to the results of a sub-set of PVS critical competencies related to this NPS element.

It could therefore also be considered to refine and group critical competencies to allow a more direct correlation of PVS results and costs for key elements of the NPS.

Governments to Keep a Record of PVS staff

OIE member countries should collect data on staff numbers of the public Veterinary Services across all levels of government.

Although collection of such data would require additional efforts by member governments, this would hugely improve the basis for any future economic assessment of the National Prevention System, as staff costs account for up to three-quarters of NPS operating expenditures in the case study countries. This could be encouraged by revising the reporting format for the annual OIE World Animal Health Report.

A possible reporting format, suggested in this study, would differentiate between public and private veterinary personnel, differentiate the categories of veterinary personnel paid from the public budget and differentiate the type of activity of the personnel.

Standards should be Established for Future Comparisons

A 'gold standard' or quality benchmark figures are needed from the OIE for comparison of NPS expenditures between countries, but assessments may be more effective if focused on key elements rather than on the total NPS expenditure at national level.

The authors says that the results of their study suggest a gradual approach to derive benchmark values that provide guidance to countries for allocating their NPS expenditures effectively and efficiently, focusing on key elements of the National Prevention System (such as cost of surveillance, border inspection, diagnostic laboratory facilities).

Benchmarking of Costs would Aid Budgeting

Finally, the authors recommend that consideration should be given to the development of a database of benchmark cost data concerning specific components of NPS expenditures. The necessary data could be obtained during the PVS Evaluation or PVS Gap Analysis visit or, alternatively, through a visit of a specialist expert team.

Benchmark cost data concerning key elements of the NPS would create a better basis for the design and budgeting of desired improvements in the NPS provisions in developing and transition countries, creating both a better basis for the budgeting process of specific countries and more transparency for donors.

Animal Cloning and Implications for the Food Chain

This report examines the potential issues involved with cloning animals for commercial industry, prepared for COI, on behalf of their client, The Food Standards Agency.

Executive Summary

Background

Animal cloning is an emerging technology in the EU (although already more established in the US), and there is potential that, if its use becomes economically viable, food derived from cloned animals will enter the food chain across the world.

It is thought that this topic is likely to encounter a significant amount of consumer interest as the technology develops. For this reason, the Food Standards Agency commissioned research to explore initial public perceptions of animal cloning and to identify what the key issues and areas of concern/uncertainty are, particularly in relation to food.

Research Objectives

The overall aims of the research were to inorm the development of the Agency's communication around this issue, to ensure that the consumer is fully informed of all aspects of the technology and to ensure that all areas of potential public concern are addressed when the acceptability of cloned animals for food production is being assessed in the EU and therefore becomes an issue more visible to the general public.

The research objectives addressed the following key areas:

1. Perceptions of current farming practices including breeding practices and views on cloning as an assisted reproductive technology
2. Levels of knowledge and perceived benefits of animal cloning
3. Animal welfare issues and other ethical concerns
4. Safety concerns in relation to food derived from clones and their offspring
5. Views on the need for, and nature of, regulation of animal cloning
6. Views on labelling of foodstuffs linked to cloned animals
7. Views on the role of the FSA in relation to this new technology.

Research Method

Given the complex nature of the topic and the evidence from previous research that public awareness and understanding of the issues are limited, a deliberative approach was adopted based on reconvened workshops, with participants taking part in two three hour sessions as well as carrying out their own background reading and research. The first workshop focused on current livestock breeding methods, an explanation of how clones are produced, how this technique can be applied to animal livestock breeding and the implications of this for the food chain. The second workshop focused on participants' views on buying and eating food derived from clones and their offspring as well as the steps they thought should be taken if such food went on sale in the UK.

Four sets of workshops were conducted, one in each of England, Scotland, Wales and Northern Ireland.

Key Findings

Current Animal Livestock Breeding

- It was accepted that livestock breeders actively manage the process in order to ensure they breed from their 'best' animals. With the exception of artificial insemination, most people had not heard of the various forms of assisted reproductive technologies currently in use
- In this context, most participants felt animal cloning represents a quantum leap from 'giving mother nature a helping hand' to 'interfering with mother nature'.

Understanding of Cloning and Perceived Benefits

- Initial levels of knowledge about, and understanding of, cloning varied widely.
- Participants struggled to identify any convincing benefits of the technique. They felt the only 'winners' were likely to be biotech companies, livestock breeders, farmers or food retailers and they were concerned that the main motive for introducing animal cloning was a pecuniary one. They questioned whether consumers would derive any tangible benefits.

Animal Welfare and Other Ethical Issues

- As participants learned about the current low efficiency rates of the cloning method they became increasingly concerned about the implications for animal welfare. This became a significant factor behind their reluctance to accept food derived from clones and their offspring.
- The research highlighted a number of other concerns that the public are likely to voice in relation to the use of animal cloning for food production. These included concerns about where the technology might lead (in particular, to human cloning) and whether mankind has the moral right to pursue such a course.
- Underpinning many of their concerns was a lack of trust in the various players involved including biotech companies, scientists, livestock breeders, farmers, government, food manufacturers and retailers.
- If the FSA, or any other body, wishes to be a credible and reliable source of independent advice in this area, it is essential that it

is seen to transcend the needs and aspirations of these different players.

Safety Concerns in Relation to Food Derived from Clones

- Opinions were shaped by previous events especially in relation to BSE/vCJD and GM food.
- Many participants were concerned that cloning could result in food that was unsafe for human consumption. This was partly a function of the perceived high incidence of miscarriages and deformed and short-lived offspring resulting from the process. It was also because of a fear that the process of cloning might somehow create new diseases or affect the food in some way that will be harmful to humans.
- There were also concerns that cloning might impact on food quality, consistency, uniformity and price.
- There is a major mismatch between the methods used by regulatory authorities to assess food safety and the public's perception of what is needed. Participants wanted to see methods for assessing food safety that were analogous to the approach used in clinical drugs trials.
- If the efficiency of cloning can be greatly improved, this will lessen the idea that the resulting offspring may pose food safety concerns. However, unless the mismatch in perceptions about the required method of assessing food safety can be addressed, the public are likely to harbour major concerns that such food is unsafe to eat.

The Need for Regulation of Animal Cloning

- If food derived from clones and their offspring were to go on sale in the UK, the research has provided a clear steer in terms of the steps that would help to increase consumer confidence. Irrespective of how participants felt about buying and eating such food, there was a high level of agreement about how it should be introduced and regulated.

This included:

- o regulations that address the entire process from animal breeding and welfare to food production and human health, including the import and export of clones, their offspring and semen/embryos, and food derived from such
- o some form of licensing not only of the process of cloning animals but also covering how such animals enter the food chain

- o an agreed set of standards and procedures coupled with proactive monitoring and enforcement
- o traceability of clones and their offspring.

- Informing and educating the public about current regulations may help increase consumer confidence when coupled with (possibly) new controls on who can clone, how they do it and how such animals end up in the food chain.

Views on Labelling

- There was a call for all food derived from cloned animals and their offspring to be clearly labelled-not just from a food safety perspective but to enable consumers to make an informed choice. The greatest challenge lies in working out how far removed an animal needs to be from a cloned ancestor before it is considered 'normal'.

Role of the Food Standards Agency

- The FSA – possibly in partnership with other bodies-was seen by most as having a key role to play in the debate about food derived from clones and their offspring both in terms of setting and policing the rules as well as informing and educating the public.
- Whatever its role, it is crucial that it is perceived to be independent and trustworthy.

Gender Differences

- There was evidence of a clear gender divide. Men often took a more rational approach, were somewhat less concerned about, and more willing to consider buying and eating, food derived from clones and their offspring. Women seemed to engage at a more emotional level, often as mothers/grandmothers, and were more worried about animal welfare and food safety. As a result, they were more likely to reject the idea of buying/eating such food. Given the fact that women tend to be the main food shoppers in many households, their views on such food are likely to have the greatest impact on any future uptake.

The key areas of concern that participants expressed are summarised below:

Based on this research, if the general public are to accept the idea of buying and eating food derived from clones and their offspring, each of these concerns would need to be addressed.

The Use of Animals in Biomedical Research: Improving Human and Animal Health

Animals used in biomedical research help us:

- understand how our bodies work
- find cures and treatments for diseases
- test new drugs for safety
- evaluate medical procedures before they are used on people.

Why Do We have to Use Animals in Research?

Although animals appear to be very different from us, their bodies work in many of the same ways ours do. Researchers who study animals discover information that can't be learned from other sources. Biomedical research saves lives. Research using animals improves the lives of millions of people each year by giving doctors clues to prevent, treat, and cure illnesses like cancer, diabetes, and AIDS. New drugs and treatments for diseases are tested on animals to make sure they are safe for people to use.

What Kinds of Animals are Used in Research?

Laboratory mice are used more often in research every year than any other animal species. Mice, and other rodents such as rats and hamsters, make up over 90% of the animals used in biomedical research. In addition to having bodies that work similar to humans and other animals, rodents are small in size, easy to handle, relatively inexpensive to buy and keep, and produce many offspring in a short period of time.

However, rodents may not always be the best animal model to use in certain experiments. In these cases, dogs, cats, rabbits, sheep, pigs, fish, frogs, birds, nonhuman primates, or other kinds of animals may be used. All of these animals together make up less than 10% of the animals used in research.

Why Can't We Use other Methods?

We use other methods whenever we can. Computer models and cell tissue studies are used in addition to animal research to discover new ways of solving complex problems. By studying cells in test tubes, scientists can learn much about how our bodies work, react to disease, and respond to treatment. Computer simulations also aid researchers. But cells and computer models simply cannot mimic the complexities of our bodies. And that's where animals come in.

Because of biomedical research we have:

- heart by-pass and other life-saving surgeries
- organ transplantation

- vaccines to prevent childhood diseases
- many other treatments and cures for diseases.

Together for Life

In the last century, most medical discoveries were possible because of animal research. To build on the great progress we've made in understanding and treating diseases, we need to continue these studies.

Cures for diseases such as cancer, AIDS, Alzheimer's disease, diabetes, cystic fibrosis, and muscular dystrophy are in reach, thanks to the animals used in research. Only through the humane use of animals in research can we hope to continue to improve the lives of both animals and humans. We all owe them a debt of gratitude for the part they play in saving the lives of millions of people around the world.

The Judicious Use of Antimicrobials for Animal Producers

To achieve that goal, producers should be committed to the practice of disease prevention through the use of vaccines, parasite control, stress reduction, environmental management and proper nutritional management. Responsible and timely management practices can reduce the incidence of disease and therefore reduce the need for antimicrobials; however, antimicrobials remain a necessary tool to manage infectious disease in beef herds.

Prudent use of antimicrobials is important to reduce livestock pain and suffering as well as minimise losses due to disease. Furthermore responsible antimicrobial use will help minimise antimicrobial resistance in bacteria, which can impact animal health in your operation and the safety of food you produce.

Like other species, bacteria find ways to survive. Whenever bacteria are exposed to antimicrobials, some survive by developing resistance. Resistance depends upon the type and amount of antimicrobial used, the number of animals treated, and the length of treatment. Therefore, it is vital to use antimicrobials appropriately and judiciously in animals under your care.

Antimicrobial resistance concerns are shared by producers, veterinarians, governmental agencies, the public health community and consumers. Of particular concern is the risk of developing antimicrobial resistant bacteria in animals and the potential for subsequent human exposure. The antimicrobial resistant bacteria can, for example, be consumed by humans along with the meat product. Producers must be committed to appropriate and judicious antibiotic use in order to (1) minimise the risk of antibiotic resistant bacteria in animals, (2) maintain the long-term effectiveness of antimicrobials for humans and livestock, and (3) protect future antimicrobial availability.

Judicious Antimicrobial Use Guidelines

Producers do not stand alone in addressing this issue. Various veterinary groups have developed and adopted judicious antimicrobial use guidelines. These guidelines provide a road map for proper antimicrobial therapeutic use in livestock.

Prevent Disease

- Prevent disease through management by emphasizing animal husbandry, biosecurity, and health maintenance. Antibiotic use cannot replace sound management practices.
- Appropriate and timely management practices include:
    - o Vaccinations
    - o Parasite control
    - o Stress reduction
    - o Nutritional management
    - o Environmental management.

Diagnose Sick Animals

- Sick animals indicate a breakdown in preventative practices.
    - o Diagnose problems early and accurately.
    - o Many diseases have similar symptoms/signs.
    - o Not all diseases are treatable with antimicrobials.
    - o Use your veterinarian and a laboratory to help diagnose the problem.

Select Appropriate Antibiotics

- Use veterinary advice and laboratory results to help select appropriate antibiotics or treatment alternatives.
- Select and use antimicrobials carefully and only when justified.
- Develop written treatment protocols in collaboration with your veterinarian.
- Use the proper dose, given by the proper route for the proper length of time.
- Treatment programs should reflect best use principles.
- Treat the fewest number of animals possible.
- Withhold treated animals or animal products properly.

- Antimicrobial use, other than according to the label and, outside of a valid veterinarian-client patient-relationship (VCPR), is a violation of federal law.

Keep Records

- Record animal or group identification, drug used, date treated, dosage used, route and location for administration, and who administered the product.
- Review records before marketing to ensure proper meat withdrawal.
- Keep and review records.
- "You cannot manage what you do not measure."

How should this Work on Your Operation?

Have a VCPR

One of the basic principles of proper antibiotic usage is having a working relationship with your veterinarian, commonly called a VCPR. This means your veterinarian knows your operation, your management, your livestock, and is involved in diagnosis and treatment decisions.

Your veterinarian must be available for follow-up in case of adverse reactions or treatment failure. This also means that you are willing to follow his/her instructions for antibiotic usage. A VCPR is not just having someone write a prescription, sell you drugs, or make them available to you.

The law requires a valid VCPR if an antibiotic is administered in any manner other than expressly stated on the label or the antibiotic is a prescription product. It is a good idea to get your veterinarian's advice on the use of any drug in your operation.

A Valid Veterinarian-client-patient Relationship is One in Which

A veterinarian has assumed the responsibility for making medical judgments regarding the health of (an) animal(s) and the need for medical treatment, and the client (the owner of the animal or animals or other caretaker) has agreed to follow the instructions of the veterinarian; (2) There is sufficient knowledge of the animal(s) by the veterinarian to initiate at least a general or preliminary diagnosis of the medical condition of the animal(s); and (3) The practicing veterinarian is readily available for follow-up in case of adverse reactions or failure of the regimen of therapy. Such a relationship can exist only when the veterinarian has recently seen and is personally acquainted with the keeping and care of the animal(s) by virtue of examination of the animal(s), and/or by medically appropriate and timely visits to the premises where the animal(s) are kept.

Establish Written Protocols

A written protocol serves as a guide for on-farm diagnosis and treatment decisions. A complete protocol should include signs of the disease and detailed directions for treatment, including meat withdrawal times. You should establish written protocols for any antimicrobial use on your operation. These protocols must include your veterinarian's input if prescription drugs are used or if drugs are used other than according to the label (extralabel use).

Extralabel Drug Use (ELDU)

Extralabel use means actual use or intended use of a drug in an animal in a manner that is not in accordance with the approved labelling. This includes, but is not limited to, use in species not listed in the labelling, use for indications (disease or other conditions) not listed in the labelling, use at dosage levels, frequencies, or routes of administration other than those stated in the labelling, and deviation from the labelled withdrawal time based on these different uses.

Understand What Constitutes Extralabel Drug Use (ELDU)

Drugs are approved by the FDA for treatment of specific diseases or conditions in a specific species or class of animal, at specific dose(s), route(s), duration(s), and frequency(s) of administration with a corresponding slaughter withdrawal time. Any use that deviates from the label instructions constitutes extralabel drug use and requires a valid VCPR.

An example of extralabel drug use is buying procaine penicillin G at a farm store and using it in cattle at a dose higher than specified on the label. The label dosage is 1 cc(ml) per 100 lb of body weight, but it is commonly used at higher dosages and/or given under the skin rather than in the muscle.

These are extralabel uses, and it is important to understand that non-prescription drugs require a valid VCPR and appropriate labelling for extralabel use. Because approved withdrawal times are based on label directions, any other use may result in residues in violation of federal law.

Prohibited Extralabel Uses

The following drugs are prohibited for extralabel use in food-producing animals (current as of May 2002)

a. Chloramphenicol;
b. Clenbuterol;
c. Diethylstilbestrol (DES);
d. Dimetridazole;
e. Ipronidazole;
f. Other nitroimidazoles;
g. Furazolidone;
h. Nitrofurazone;
i. Sulfonamide drugs in lactating dairy cattle (except approved use of sulfadimethoxine, sulfabromomethazine and sulfaethoxypyridazine);
j. Fluoroquinolones; and
k. Glycopeptides.

An example of an illegal extralabel drug usage would be the administration of fluoroquinolones (e.g. enrofloxacin) to treat calf diarrhea.

It is illegal to use any feed additive drug in an extralabel manner in beef cattle.

Train People Who Treat Livestock on Your Operation

People responsible for treating livestock must be properly trained to recognise common signs of disease as well as proper drug use and record keeping. Periodic and timely review of written protocols and treatment records with your veterinarian is the most effective method for ensuring judicious antibiotic use.

Other sources of information regarding proper use of antibiotics include extension specialists, producer associations and pharmaceutical technical service representatives. However, recommendations for extralabel usage can only be obtained from your veterinarian.

Foot and Mouth Disease: Novel Technologies Improve Detection and Control

Scientists at the USDA Agricultural Research Service (ARS) are using infrared technology to identify cattle infected with foot and mouth disease (FMD) virus. The same group has developed the world's first molecular-based FMD vaccine, which has been effective in pigs, writes Laura McGinnis, ARS staff.

Infrared thermography (IRT) cameras can see what human eyes cannot: heat. That is why scientists at the ARS Plum Island Animal

Disease Centre (PIADC) at Orient Point, New York, are using the technology to identify cattle that may have been infected with foot-and-mouth disease (FMD).

"This IRT technology has been used for years and has applications for a variety of fields, from astronomy to law enforcement to the military," says PIADC research leader, Luis Rodriguez. "This is the first scientific report of IRT as a tool for early detection of FMD-infected animals."

Chemist Marvin Grubman and visiting scientist Fayna Diaz-San Segundo examine a plate used in a test to determine the correct amount of a new foot-and-mouth disease vaccine to administer to cattle.

The United States has not had an outbreak of FMD since 1929 but there is no way to guarantee that this very contagious disease is gone forever – as the United Kingdom learned in 2001, when an outbreak of FMD ended a 34-year disease-free streak. The disease spreads rapidly and wreaks havoc on trade and transportation, so being prepared for an outbreak is a priority for the US government.

An essential component of containment and eradication is the ability to quickly assess the scope of an outbreak.

"It's a huge endeavour to examine the animals one by one. Even on a small farm it's a big task," says Dr Rodriguez. He and his colleagues evaluated a new method for rapidly and accurately detecting animals at risk of infection. Their method, which uses IRT cameras to identify potentially infected animals quickly, is not intended to be a diagnostic test. Rather, it enables scientists to concentrate their resources by quickly isolating animals that require further testing with a disease-specific method.

IRT cameras can identify at-risk cattle 48 hours before they begin to show any clinical symptoms. In the event of an outbreak, this technology could facilitate rapid containment of the disease.

Spotting Hot Hooves

Infrared image of a cow not infected with foot-and-mouth disease virus (above). Note that the hooves are not red. Red colour in the hooves would indicate heat.

Infrared image of a cow infected with foot-and-mouth-disease virus. Note that the hooves are red. Red colour in the hooves indicates heat.

How does it work? Objects of all temperatures above absolute zero (about minus 460°F) emit infrared radiation. The hotter the object, the higher the emissions, and IRT cameras make those emissions visible.

Foot temperatures rise in cattle infected with the FMD virus, resulting in a visible difference in IRT photographs. This phenomenon was observed by Craig Packer, a wildlife biologist at the University of Minnesota. While tracking migratory animal populations in Africa for an unrelated study, Dr Packer used an infrared camera to take photographs of wild mammals and observed that some impala had very hot heads and feet. These animals, he learned, were suspected of carrying FMD. Curious about the implications, Dr Packer contacted the scientists at PIADC.

They followed up on Dr Packer's tip and conducted a formal study in PIADC's secure Biosafety Level 3 facility. Their results showed that IRT photography could detect elevated hoof temperatures up to two days before cattle developed clinical signs.

"By looking at the hottest spot in the hooves, we could identify FMD infected cattle when that temperature was higher than 34.4°C," Dr Rodriguez says. "It allowed us to predict with up to 88 percent confidence which animals would develop symptoms within 48 hours."

In an IRT photograph of a group of cattle, the hooves of healthy animals appear blue-green, whereas infected cattle have orange-red feet. This easy visual distinction could allow scientists and veterinarians to identify potentially infected cattle in large groups without examining animals individually. The technology is cheaper and faster than existing screening methods, which involve individual clinical examinations for every animal. IRT can be paired with other ARS-developed tools for superior disease-outbreak response. One such tool is a rapid diagnostic test Plum Island scientists developed in response to a 2002 request from the US Congress.

The PIADC scientists collaborated with the biotechnology company Tetracore to develop the test, which can detect RNA from the FMD virus in less than two hours. It uses technology known as 'real-time PCR'. USDA's Animal and Plant Health Inspection Service has adopted the test and uses it in the National Animal Health Laboratory Network. In the event of an FMD emergency, laboratories throughout the United States could use the test to diagnose samples rapidly. It can also be used to quickly distinguish between FMD and vesicular stomatitis, an animal disease with similar symptoms that occasionally crops up in the United States.

A Safer, Better Vaccine

Another important FMD development from PIADC is the world's first effective molecular-based FMD vaccine for cattle, developed by chemist, Marvin Grubman, and colleagues. The vaccine can be produced without using infectious FMD materials, which means it can be produced safely, in the United States, without expensive, high-containment facilities.

Another benefit is that animals vaccinated with Dr Grubman's vaccine do not produce all the antibodies that infected animals produce. Because of this, they can easily be distinguished. Most existing vaccines cause animals to produce the same antibodies as infected animals, making it difficult to determine whether an animal with antibodies was naturally infected or vaccinated.

At the ARS Plum Island Animal Disease Centre in Orient Point, New York, wildlife ecologist Craig Packer (left) and microbiologist Luis Rodriguez use an infrared thermography camera to identify cattle infected with foot and mouth disease virus.

ARS – in collaboration with the US Department of Homeland Security's Targeted Advanced Development unit and GenVec, Inc., a biopharmaceutical company based in Gaithersburg, Maryland – is developing this technology for potential commercialisation.

Tests have shown that the vaccine becomes effective a mere seven days after it has been administered. Although this is one of the fastet vaccines available, Dr Grubman and his colleagues wanted faster protection. After all, a lot can happen in seven days, particularly during an outbreak. In a recent study, Dr Grubman found that interferons – proteins produced by the immune system – can offer protection during that first week.

Interferons are so named because they interfere with virus replication. Dr Grubman and his colleagues introduced type I and type II interferon genes into an adenovirus, which they administered to groups of pigs. One day later, they infected the swine with FMD virus. The interferons blocked FMD virus replication, offering the swine protection. "The interferon gives early protection for three to five days while the animals are developing an antibody response to the vaccine," Dr Grubman says. "This significantly increases their chances of resisting FMD."

ARS and GenVec are now collaborating to combine the interferons and the FMD vaccine so they can be administered together.

Health and Welfare of Cattle Transported in Late Pregnancy

.Since July 2009 the Veterinary Laboratories Agency (VLA) has investigated the deaths of dairy heifers transported from Europe in late pregnancy. They died from severe metabolic disease within six weeks of transport, with fatty liver being the most common feature.

Further cases have included death and dehydration of pregnant heifers during transport. Metabolic disease only occurs in certain batches of heifers transported in late pregnancy, possibly associated with animals in very good or over-fat condition at calving. However, while thousands of cattle are regularly transported without apparent incident, it is recommended that pregnant cattle, and in particular heifers, should be carefully managed before, during and after long journeys.

Since 2008 Animal Health has reported a number of incidents of cattle calving during or just after transportation, contrary to current EU regulations.

There are significant risks to the health and welfare of heifers and cows transported over long distances in late pregnancy. The risk may be greater in heifers because they tend to have a shorter pregnancy and because more of them are transported when pregnant (the number of imported dairy heifers has increased in the past two years). Additionally, since most pregnant heifers are physically immature during pregnancy they may be more susceptible to the stress of long distance transport.

Farmers (both vendors and purchasers) and hauliers must be aware of these risks, take action to mitigate them, and be aware of their statutory obligations. Council Regulation EC/1/2005 lays down the provisions on the protection of animals during transport.

This is enforced in England under The Welfare of Animals (Transport) (England) Order 2006, and states that pregnant females should NOT be transported when 90 per cent or more of the expected pregnancy period has passed, or in the week after calving.

Important Factors to Consider When Transporting Pregnant Cattle

Energy and water requirements increase considerably in late pregnancy. Evidence indicates some heifers experience significant negative energy balance during transport, while others suffer from dehydration.

There may be an interruption in food and water intake (even if it is provided to the animals) during transport, particularly on long journeys.

A significant change in dietary management in late pregnancy is more likely to lead to metabolic disease. In some cases, heifers reared indoors at the source farm were put out to grass at the destination farm.

The stess of transportation may intensify these problems and lower animals' resistance to disease.

The Law on Transportation

It is an offence to transport any animal in a way which causes it, or is likely to cause it, injury or unnecessary suffering. Transporters and keepers can be prosecuted if they flout the law. In addition, animals which are unfit must not be transported. This includes pregnant dairy cattle transported in the final 10 per cent of their gestation period (with the exception of transportation for veterinary treatment), and in the week following calving.

In addition, under the Animal Welfare Act 2006, animal owners and keepers of animals have a duty of care to provide for their basic needs and to prevent suffering. Someone responsible for an animal, who permits another person to cause unnecessary suffering, is also committing an offence.

This is also the case if they fail to take reasonable steps to prevent suffering during and after transport. Intra-Community Trade Animal Health Certificates clearly state the requirements for compliance with the European Transport Regulation EC/1/2005.

Best Practice and Recommended Actions

Whilst there are legal limitations regarding the transportation of pregnant cattle in the final 10 per cent of gestation, there are increased risks to cattle health and welfare throughout pregnancy. Therefore the journey must be planned and managed to reduce these risks and to ensure that their metabolic requirements will be fully met. Check exact insemination or service date and expected calving date for each animal in the batch.

Avoid negative energy balance in late pregnancy by ensuring appropriate feeding before, during and after transport.

Avoid sudden changes of management and diet for pregnant animals. Upon arrival at the farm of destination, heifers should initially remain on the same diet as before transportation and new foods should be introduced gradually.

It is recommended that pregnant cattle are not transported over long distances in the last two months of pregnancy. This will also allow the animal to adapt to new rearing conditions before calving.

Always consider the disease status of new animals introduced in any herd. This is especially important if the animals originate from abroad because they may carry diseases that are not already present in GB. The biosecurity of the herd of destination should be managed through a herd health plan.

Cold Stress in Cows: What iCold to a Cow?

When temperatures start to decline in winter, particularly as we get closer to 0°C (or 32°F), it is time to think about what effect this is having on cow productivity and efficiency.

Like all mammals, cows are warm blooded and need to maintain a constant core body temperature. Normal rectal temperature for a cow is around 38°C (101°F).

Within a range of environmental temperatures called the "thermoneutral zone," animals do not have to expend any extra energy to maintain their body temperature. At the lower end of this range, normal metabolic processes supply enough heat to maintain body core temperature. Within their thermoneutral zone, animals may modify their behaviour, such as seeking shelter from wind, and respond over the long term by growing a thick hair coat for winter, without affecting their nutrient requirements.

However, below the lower limit of the thermoneutral zone, in the "lower critical temperature," the animal experiences cold stress. To combat cold stress, the animal must increase its metabolic rate to supply more body heat. This increases dietary requirements, particularly for energy. Typical lower critical temperatures for beef cattle are affected by a number of factors. Table 1, below, shows the impact that different hair coat types can have on lower critical temperature.

Cattle, like humans, actually experience the "effective temperature," which takes into account both air temperature and the effect of wind chill. Cool or cold wind passing over an animal draws heat away from it much more quickly than still air at the same temperature. Wind chill effects for cattle are shown in Table 2, below. These figures assume a dry, clean hair coat. If the animal is wet and/or dirty, consider the data to be an underestimation of the effect of the wind.

Assumes that hair coat is dry and clean.

For example, when air temperature is –18°C and wind speed is 24 kph, the effective temperature experienced by the animal is the equivalent of a still air temperature of –26°C.

If cows are exposed to wind or drafts, it is important to adjust for the effective temperature and take the appropriate steps to ensure that the cows can maintain body temperature and weight.

Factors Affecting an Animal's Ability to Withstand the Cold

- Acclimation: Cattle do adjust or acclimate to colder weather by growing a longer, thicker coat. This provides additional insulation against cold weather. The coat must be clean and dry to provide maximum protection to the cow. Dirt or moisture on the coat reduces its inulation value dramatically.

- Fat Layer: Cattle in good condition with a thick fat layer are better able to withstand the cold than thin cattle. The fat layer acts as another insulating layer between the animal's core and the environment.
- Metabolic Rate: Cows will also increase their metabolic rate to increase heat production and help maintain body temperature. This increases the need for dietary energy, so appetite is usually increased and cows eat more.

The Effects of Severe Cold Stress on Cattle

Hypothermia occurs when the body temperature drops well below normal. In general terms, with cattle, mild hypothermia occurs with a body temperature of 30°C–32°C, (86°F–89°F), moderate hypothermia at 22°F–29°C, (71°F–85°F) and severe hypothermia below 20°C (68°F). As rectal temperature drops below 28°C (82°F), cows are not able to return to normal temperature without assistance through warming and the administration of warm fluids. As hypothermia progresses, metabolic and physiological processes slow down, and blood is diverted from the extremities to protect the vital organs. Teats, ears and testes are prone to frostbite. In extremes, respiration and heart rate drop, animals lose consciousness and die.

In most situations, a more insidious and costly problem occurs. Cows are subjected to an environmental temperature below the lower critical temperature, but without obvious signs of hypothermia. This increases the maintenance energy requirement of these animals as they adjust to the conditions and divert more energy to maintaining body temperature.

There are two potential responses to this situation.

Cows have access to higher quality feed and/or increased intake, and therefore maintain their body weight.

Cows try to increase feed intake in an effort to meet their energy requirements. Given the opportunity and gut capacity, cows will eat more feed to help meet their increased energy demands. Practically, it is usually expedient to feed grain as well. This increases feed costs, increasing the cost of keeping cows, however the expectation is that cows will maintain their body weight!

It is generally accepted that for every 1°C drop below the lower critical temperature, there is an approximately 2% increase in energy requirements. The amounts of additional feed required for a cow under cold stress can be calculated, but as a rule of thumb, a cow with a dry winter coat should be fed the additional feed as presented.

Cows may not be able to eat the amount of extra hay required to maintain their body weight and may have to be fed the indicated amount of grain instead of additional hay to meet their energy requirements.

Cows don't have increased feed quality and intake and lose body weight.

If cows are not fed additional feed or the quality does not allow them to eat enough to meet their additional energy requirements, body mass will be "burned" to produce metabolic heat. These cows lose weight as both feed energy and stored fat are diverted to maintain body temperature and vital functions.

Cows in this situation that start to lose weight soon enter a downward spiral — the more weight (fat) they lose, the less insulation they have, the more susceptible they are to further cold stress, and they lose weight even faster.

Cows, and especially heifers that lose weight, calve in poor condition. The consequences are increased calving difficulties, an increase in the number of lighter, weak calves and higher calf mortality. These dams produce a reduced amount of colostrum (of lower quality) and have lower milk production, increased neonatal mortality and reduced growth rate in surviving calves. These cows usually have delayed return to estrus, longer days open and poorer reproductive success.

Key management factors to limit the effects of cold stress :

- Monitor the weather. Monitor temperature and increase feeding in response to cold weather. Cows in the last trimester require additional grain feeding during periods when the effective temperature falls below the lower critical level.
- Protect animals from the wind. Wind markedly reduces the effective temperature, increasing cold stress on animals.
- Bed cows well. Providing adequate dry bedding makes a significant difference in the ability of cattle to withstand cold stress.
- Keep cows clean and dry. Wet coats have greatly reduced insulating properties and make cows more susceptible to cold stress. Mud-caked coats also reduce the insulating properties of the hair.
- Provide additional feed. Feed more hay and grain. If wet feeds are fed, make sure they are not frozen.

- Provide water. Make sure cows have ample water available at all times. Limiting water will limit feed intake and make it more difficult for cows to meet their energy requirements. Frozen troughs and excessively cold water seriously limit water intake.

We can't control the weather but we can do everything reasonably possible to reduce the effects of cold on cows. This will help reduce costs and improve production efficiency.

Nursing Calf Deworming

Carter and Matthew J. Hersom· Published by the University of Florida IFAS Extension. Historically, the predominant health problem worldwide for ruminant animals has been the presence of internal parasites.

The continued use of anthelmintics, or dewormers, remains controversial among researchers because there is wide variation among results when these products are used on cattle with moderate to low levels of parasite infestation. It is clear, however, that with any level of infestation above moderate, dewormers provide almost immediate responses and improvement in animal health and performance.

Parasitic infection and disease receives relatively little attention in most areas. Likely, we are so accustomed to treating our cattle on a routine basis that clinical signs of disease are rarely observed. Immunity to parasites increases with age. Older animals like mature cows in particular have the ability to ward off many parasitic challenges, or at least keep these invaders to a minimum.

Parasites, however, may be silent rustlers of performance while existing in a sub-clinical fashion in your herd. Although these infections may not be significant enough to manifest clinical signs, weaning weights may be improved when nursing calves are dewormed at branding, or approximately three months of age. In fact, deworming calves may be the most profitable task you can perform.

If you think the effect of any given parasite stops simply with lower weight gains and an unthrifty appearance, you may be misinformed. Parasites can contribute to other conditions like anemia which can be significant because of its affect on red blood cells and thus, oxygen transport. They may also affect the immune system's ability to respond to vaccines by producing lower-than-normal white-cell populations within the body.

Beef cattle are susceptible to various parasites. Here are a few species that are important. Check with your local veterinarian for an appropriate deworming program for your herd. Be sure that you are aware of liver flukes in your area and whether or not your herd is exposed to these, as well. Not all dewormers are labelled for the control of flukes, so check your products carefully and again, consult your veterinarian.

Gastrointestinal Roundworms

- Ostertagia ostertagi
- Cooperia oncophora
- Haemonchus placei
- Trichostrongylus spp.
- Strongyloides papillosus.

Lungworms—Dictyocaulus Viviparous

Standard procedure for parasite control at most beef operations in the U.S. is to treat beef cows once or twice annually and possibly to deworm the calf at weaning only. Recent field trials have dismissed the dogma that calves did not have a sufficiently high level of parasitism to warrant treatment until weaning, or after. Industry-funded research has recently demonstrated higher weaning weights of calves treated with doramectin (Dectomax™, Pfizer Animal Health, New York) prior to weaning. Their data indicated a 25 lb advantage in weaning weight resulting in additional profit per head ($17.65), as well as a nearly four-fold return on investment (ROI) over the cost of the deworming product alone.

Deworming Calves Still on the Cow

Spring born calves (n=567) from three geographically different locations within the University of Florida/IFAS system were utilised to conduct a nursing calf deworming experiment: NFREC (MAR; n=177), Marianna, FL; Beef Research Unit (BRU; n=186), Gainesville, FL; and Boston Farm – Santa Fe River Ranch (SF; n=204), Santa Fe, FL. At least two breed types were available at each location, including Angus, Brangus, Brahman, and Romosinuano, as well as some graded combinations of these breeds (composites).

Although processing calendar dates varied by location, the project began after June 1, 2005 and depended on the projected weaning date determined by each unit manager. Treatments included a control group (CONT), which received no deworming compounds during the study, and a treatment group (DW) which was dewormed with injectable doramectin (1 mL per 110 lb BW, subcutaneous [SC]) 90 days prior to the projected weaning date. Calves were blocked by site and randomly assigned to either CONT or DW; equal numbers were assigned to treatments as much as possible and necessary to balance the experimental design.

On day 0 all calves were individually weighed in order to get an accurate body weight which was used to calculate an appropriate dosing rate for DW. After weighing on day 0 calves were returned to their dams and taken to designated pastures for grazing. Cow-calf pairs within each unit and among treatment groups were grazed on similar forage types and received similar nutrition at all times in order to avoid bias based on forage or nutrition limitations. Treatment groups were intermingled as necessary depending on pasture conditions and overall grazing logistics. Dams were body condition scored (BCS) on day 0 of the study and at weaning to determine if any change in calf growth rate caused by the treatments may have had an indirect effect on the physical condition of the dam. Calf weights and dam BCS were also obtained at a midpoint during the study.

Pasture forage allowance was measured as cow-calf pairs were introduced to new grazing sites and when they were removed in order to estimate forage consumption during grazing.

Results

Performance data from this study are presented in Table 1 and Table 2. Across all locations, DW calves gained more total weight and ADG was greater among DW calves compared with CONT calves. On average in this experiment, deworming cost approximately $1.57 per head. DW calves returned $9.57 per head more net revenue ([lb total wt. gain x $1.28/lb BW average sale price of calves]-$1.57/hd deworming cost) when considering only the cost of the deworming product. Labour costs should be considered.

Implications

Under these experimental conditions, these data indicated both an animal performance advantage and a positive ROI. Given these results and the only modest economic improvement, economies of scale may limit the acceptance of this process to larger operations. The cost:benefit ratio may not be as significant for the average producer, especially if it means putting the herd through the chute an additional time. Labour costs, if calculated at two dollars per head, could consume nearly twenty percent of the increased revenue. Dollar value gains in the range of $15-$25 would likely make this more widely attractive. Of course this equation has various components that affect the outcome directly: calf prices, calf quality, animal performance as affected by rainfall and/or forage availability, labour and processing, etc. Some economists, however, may advise serious consideration of any procedure that adds as little as one dollar to the bottom line of any enterprise.

Ultrasound Scanning Beef Cattle for Body Composition

Ultrasound technology uses sound waves to develop images of body composition. Body composition traits that can be measured include 12th to 13th rib fat thickness, rump fat thickness, ribeye area, and intramuscular fat percentage (marbling).

Each of these traits is at least moderately heritable and is significant in the determination of red meat quality and yield for individual animals.

Chapter 8

Drug Treatment

Livestock medicines are an important tool in the treatment and prevention of disease. Correct treatment methods assure the safety of food products and insure an effective response to treatment.

Consider the following points before treating dairy cattle.

Selection of Cases

Drug treatment is not always the best option for controlling animal disease. Treat animals based on the diagnosis, the expected response to treatment and the economic benefit expected. For example, viral infections do not respond to antibiotic therapy while those caused by bacteria will. Treating subclinical mastitis at dry off is effective and economical while treating cows during lactation may not be. Select suitable cases to treat with the help and advice of the veterinary practitioner.

Drug Selection

Select the correct therapy. Consult your veterinarian for advice on the correct medication, the route of treatment, the treatment dosage, the time between treatments and the number of treatments. Veterinarians should leave clear written instructions with the herd owner identifying the treated animal and giving information on the treatment protocol. The veterinarian also plays an important role in monitoring the response to treatment.

Treatment Method

Treatment must be given correctly to be effective and to prevent complications. Use the following guidelines to develop good treatment habits.

- Wash your hands before and after handling livestock medicines.
- Use proper equipment: Choose the correct syringe and needle size for the dosage and the type of injection to be given.

For intramuscular injection use a I 1/2", 16 or 18 gauge needle to insure the drug goes in the muscle and not under the skin. Before injecting, pull back on the plunger to insure the needle tip is not in a blood vessel. Select appropriate injection sites with the help of your veterinarian. Read the label for the maximum amount to be injected in one site.

For subcutaneous injection use a 1/2" to 1", 16 or 18 gauge needle. Check that the needle tip is moveable. Inject a small amount of drug to see if a "bleb" of skin starts to rise in the area of the needle tip. This will verify that the needle is under the skin and not in the muscle. Inject only in sites recommended by your veterinarian.

- inject only in clean body sites.
- Use clean equipment. Single use, sterile, disposable needles and syringes are preferred.
- Give repeated injections in different body sites.
- Before infusing antibiotics into the udder, wash and dry your hands and the teat with single use paper towels. Disinfect the teat end with the alcohol swab provided in the medication package. Avoid touching the infusion canula at the end of the treatment tube. Use only single dose infusion products in disposable syringes. Teat dip the teat after infusion of medication.

Dosage Calculation

To calculate the correct dosage you must know the weight of the animal and the dosage rate. For example, to treat a 600 kg cow with procaine penicillin at the label dosage of 2.5 mL per 100 kg of body weight once daily, inject: 600 kg/100 x 2.5 mL = 15 mL.

One milliliter (ml) and one cubic centimetre (cc) represent the same volume and are interchangeable in calculating drug dosages.

Repeat Treatments

Determine the number of treatments to be given from the product label or as recommended by the veterinarian. The duration of treatment should result in a cure without risk of relapse, yet be short enough to insure withholding times are not extended.

Withholding Times

Withholding times for milk and meat are given on product labels. A withdrawal day is a full 24 hours starting after the time of treatment. A 48 hour withdrawal time for milk is illustrated.

Errors in calculating withdrawal times of only a few hours could result in a residue violation. Label withdrawals are not accurate if products are used in an extra-label fashion, if drugs are used in combination (for example intramammary and intramuscular treatments for mastitis given at the same time), or if the treated animal is severely sick and unable to clear the drug from its body at normal rates.

In these cases you must test the milk before addition to the bulk tank to insure it is residue free.

Prevent Residues

Simple management practices will prevent contamination of milk or meat. To prevent residues:

- record all treatments given;
- mark all treated cows;
- inform all people involved in milking of treated cows;
- milk treated cows last or use separate "bypass" equipment to insure that no contaminated milk enters the milk supply;
- discard milk from all quarters of treated cows;
- discard milk from all cows calving within 30 or 42 days of dry treatment according to label directions;
- discard milk from fresh cows for the required period if dry treatment was used;
- use antibiotic test kits as needed;
- and follow label directions for all medications used. These include feed additives, medicated feeds such as calf starter, topical preparations, as well as injectable and infusion products.

Treatment Records

Many antibiotic residue violations result from failure to: identify treated cows, maintain treatment records, and use proper milk withholding times. The record system must make all staff involved in milking aware of treated cows and the period for withholding milk from sale. Identify treated cows in a manner clearly visible to the person milking. Some methods used are:

- leg bands,
- coloured tape or fluorescent hockey tape around the legs or tail, or
- paint markings on the cow's flank, rump or legs.

In larger herds identification may be colour coded to show the last day to withhold milk. In tie-stall barns where cows always occupy the same stall, coloured tape or tags attached to the milk inlet of the pipeline can identify a treated animal. Reinforce cow identification systems with a prominent chalk board or bulletin board in the milking parlour or barn entrance. Walls constructed of "white board" designed to be written on with special markers are an excellent way to create a very large bulletin board. The identity of all treated cows and the date and time of the last milking withheld should be clearly visible. Train all staff involved in milking to refer to this board immediately before each milking.

Keep a permanent, detailed treatment record for reference and management purposes. Write this in the herd health book or in the individual cow record files. This record should identify the animal, the product and dosage administered, the date of treatment and the milk withholding period. Before shipping any animal for slaughter, check this record to insure pre-slaughter treatment withholding requirements are met.

In addition, store product inserts and packaging from all livestock medicines in a file folder. This "box top file" will provide additional information if questions about previous treatments arise.

On-Farm Antibiotic Testing

The persistence of antibiotic residue in milk of treated cows may vary. This may depend on the cow, her metabolism, the medicine type, the use of a combination of medications, the dosage of medication and the method of administration. Test milk suspected of contamination using antibiotic test kits. Examples of these situations where milk contamination may occur include:

- the addition of purchased milking cows to the herd for which the treatment history is unknown,
- fresh cows purchased during their dry period who have unknown dry off dates or dry treatment histories, cows treated in an extra-label manner, cows treated with more than one product, cows which are severely ill at the time of treatment,
- cows which calve before the end of the milk withholding time following dry treatment,

- establishing the identity of a treated cow when an error in identification may have occurred, and
- the testing of milk in the bulk tank when contamination with milk containing medication may have occurred.

A variety of kits are commercially available. When selecting a test kit the user should recognise that kits vary in the type of antibiotics and amount of antibiotic they detect. No single kit can detect all commonly used antibiotics. Before selecting a test kit confirm with your Ontario Ministry of Agriculture and Food, Dairy Inspection Branch field person that the kit is comparable in sensitivity to official tests. For the penicillin group of antibiotics the "Delvotest P" is easy to read and detects these drugs at levels similar to the official test.

Milk producers who are comfortable using these tests are encouraged to purchase a test kit for on-farm use. Antibiotic testing services may also be available from veterinarians, milk processing plants, and others. Local staff of the Dairy Inspection Branch have current information on test kits and are also available to conduct a field test on request.

Disposing of Milk From Treated Cows

All unmarketable milk must be disposed of in a manner which protects the environment and keeps drug residues out of all food products. Milk from treated cows can be fed to replacement dairy calves older than 6 weeks of age without harm. Do not feed variable or abnormally large amounts of milk to calves to dispose of it. Milk from treated cows cannot be used in a sour colostrum program since antibiotics prevent normal fermentation. Do not feed milk from treated cows to bull calves or other livestock which may be sold or slaughtered before they are residue free.

Milk from treated cows can be added to a liquid manure storage, or along with straw to absorb it, to a solid manure storage. Do not add milk containing antibiotic to milk house wash water entering a septic tank or treatment trench system. Milk solids will plug the trench tile.

Summary

The ultimate responsibility for insuring a milk supply free of drug residues lies with the milk producer. Use of livestock medicines is a privilege which livestock owners cannot afford to abuse. Correct usage of livestock medicines, recording of treatments, and clearly identifying treated cows are essential practices.

Review the use of livestock medicines on your farm. Use the recommendations of this factsheet to revise procedures to insure the safety and well-being of livestock, the dairy industry and of consumers of dairy products.

Nitrate Poisoning in Cattle, Sheep and Goats

Nitrate poisoning is a condition which may affect ruminants consuming certain forages or water that contain an excessive amount of nitrate.

Causes of Nitrate Poisoning

Under normal conditions, nitrate ingested by ruminant livestock, like cattle, sheep and goats, is converted to ammonia and then bacterial protein in the rumen by bacteria. The steps of conversion in this process are as follows:

—BactProtein—Nitrate (NO_3) → Nitrite (NO_2) → Ammonia (NH_3) → Amino Acid → Protein

Nitrate is converted to nitrite faster than nitrite is converted to ammonia. Consequently, when higher than normal amounts of nitrate are consumed, an accumulation of nitrite may occur in the rumen. Nitrite then will be absorbed into the bloodstream and will convert hemoglobin to methemoglobin, which is unable to transport oxygen. Thus, when an animal dies from nitrate poisoning, it is due to a lack of oxygen.

The occurrence of nitrate poisoning is difficult to predict because nitrate levels can change rapidly in plants and the toxicity of nitrate varies greatly among livestock due to age, health status, and diets. However, concern should certainly be raised when plant growth has been less than half of normal or nitrogen application more than twice recommended.

Nitrate Levels in Plants

Plants normally take up nitrogen from the soil in the form of nitrate, regardless of the form of nitrogen fertilizer (including manure) applied. However little nitrate accumulates in plants, when growth is normal, because the plant stem and leaves rapidly convert nitrate to plant amino acids and protein. Under certain conditions, however, this balance can be disrupted so that the roots will take up nitrate faster than the plant can convert the nitrate to protein.

The nitrate-to-protein cycle in a plant is dependent on three factors:

- Adequate water

- Energy from sunlight
- A temperature conducive to rapid chemical reactions.

If any one of these factors is inadequate, the root continues to absorb nitrate at the same rate while storing it unchanged in the stalk and lower parts of the leaves. When this situation develops, nitrate accumulates. Nitrates may also accumulate in plants from excessive nitrogen fertilization, for example on fields where a large amount of manure have been applied.

Some plants are more likely to accumulate nitrate that others. Crops capable of high levels of nitrate accumulation under adverse conditions include corn, small grains, sudangrass, and sorghum. Weeds capable of nitrate accumulation include pigweed, lambs quarter, sunflower, bindweed and many others. Vegetables capable of accumulating large amounts of nitrate that are most frequently grazed include sugar beets, lettuce, cabbage, potatoes and carrots.

Nitrates in Water

Nitrates and nitrites are water soluble. They move with the water. Any nitrate added to, or produced within, the soil may be leached or washed away by moving water—either by surface run-off or ground water percolation. Nitrates are more concentrated below or near the area of waste accumulation or disposal such as manure piles, feedlots, septic tank disposal fields, cesspools, privies, etc. Excess nitrates also are more apt to be found in ground water under low areas and waterways that collect or convey. Water samples from shallow, dug, bored and driven wells more frequently contain excess nitrates than water from deeper, drilled wells. Nitrate levels generally are highest following wet periods and lowest, even down to zero nitrates, during dry periods which may cause a false sense of security. Preferably, a well should be tested immediately following a wet period.

Toxicity Variation

Ruminant livestock can tolerate a wide range of nitrate, depending on several factors. Factors making nitrate less toxic include:

- The animal can become conditioned to eat larger amounts of feed with high nitrate content if the increase is gradual.
- Healthy animals are less likely to be adversely affected than animals in poor health.
- Adequate amounts of available carbohydrates (grain) allow the animal to consume more nitrate because carbohydrates enhance the conversion process from nitrate to microbial protein.

Factors making nitrate more toxic include:

- Rapid diet changes can trigger nitrate poisoning.
- Parasitism or other conditions causing anemia will increase susceptibility.
- Nitrate in more than one diet component (e.g. water and forage).

Symptoms of Nitrate Problems

Many general symptoms such as poor appetite, weak calves, lambs or kids, abortions, poor growth, and general unthrifty conditions are frequently blamed on nitrate. These and other general problems can also be caused by a number of disease, nutritional or managerial problems. Therefore, it is essential to study the situation thoroughly before concluding that nitrate is the problem.

Simple chronic nitrate toxicity is rare. More likely, high nitrate is only one of several factors resulting in poor performance. For example, poor performance on a ration that is low in energy or lacking in essential minerals is apt to be worse if nitrate is also present. When good feeding and management practices are followed, it is very difficult to produce chronic nitrate problems.

Many factors have been suggested to explain the different results obtained in research trials studying the nitrate problem. Age, condition, and species of the animal, other chemical compounds and nutrients in the ration, and the types of nitrogen compounds in the feed or water are some of the factors that must be considered.

The most dramatic nitrate toxicity problems have occurred when hungry cattle were put on corn stalks, oat straw or weedy pasture. Under these conditions the highest nitrate feeds are fed as the total ration, and the feeding of well-balanced rations and adaptation by the animal are ignored. Sudden change to high nitrate corn silage as the main feed can cause problems. Milking cows and other animals receiving large amounts of grain are not as likely to have nitrate toxicity problems as dry cows, heifers and other animals because the milking cows are on a higher energy ration and because the high nitrate feedstuff is likely to be a smaller proportion of the total diet..

Grains and other concentrates are low in nitrate. Forages (leaves and stems) will accumulate more nitrate than grains. Because forage comprises a larger percentage of ruminants dies, high nitrate in feed is more likely to be of concern in feeding ruminants than non-ruminants. However, nitrate from feed or water can cause problems for all animals and to humans.

Taking an Accurate Sample for Analysis

If fresh-chopped forage or silage is suspected of being high in nitrates, it should be tested. Samples can be analysed by some commercial forage testing laboratories as well as by the University of Wisconsin Soil & Forage Analysis Lab (8396 Yellowstone Dr., Marshfield, WI, 54449) and the Soil & Plant Analysis Lab (8452 Mineral Point Rd., Verona, WI, 53593).

The two key steps to getting accurate nitrate analysis are:

1. To make sure the sample is representative of the feed or water that is being analysed.
2. To prevent loss of the nitrate between sampling and laboratory analysis.

Preventing Nitrate Loss in the Sample

Preventing nitrate loss in the sample before analysis is critical to getting accurate results. This is difficult in silages and fresh forages therefore it is best to take fresh silages and samples directly to the laboratory. If samples must be held or shipped, samples should be frozen or dried before shipment. Freeze samples in airtight, plastic bags for at least 24 hours and ship in insulated containers to reach the laboratory while still frozen. (DO NOT ship samples late in the week to risk getting delayed over the weekend.)

Alternatively, dry samples by spreading in a thin layer on clean paper to dry. Artificial heat is desirable, but the sample should not get above 160° F. Fresh forages will ferment in airtight bags and for this reason are usually dried. Silage samples are usually frozen because it is difficult to dry without losing volatile materials.

Water samples should be collected in special water sample bottles obtained from the laboratory or from the health agency in the area. To obtain a good sample of water from a well, let the water run until water in pipes and pressure tank has been discharged. It is comparatively easy to get a good water sample and the laboratory analysis is also more standardised. There is less variation in test results from water than feed. However, nitrate levels in water can vary considerably from month to month.

Results of analysis on the same sample of feed can vary considerably due to limitations of chemical test used. Variations of 100 parts per million in a feed are well within the range of expected errors. Results of tests on feeds containing small amounts of nitrate might appear large but could be within the normal tolerance of test used. For example, if a report on one sample of silage shows 200 parts per million of NO_3-N and another report shows 300 ppm of NO_3-N, they are for all practical purposes the same. If reported as percent, the variation does not appear as large. In the above example, the percent NO_3-N is.02 and.03.

Interpreting and Using Nitrate Reports

Results of nitrate analysis may be confusing because of the variation in methods of reporting. In the chemical analysis for nitrate, the actual element determined is the oxidised nitrogen. However, values may be reported as percent nitrate (NO_3), or nitrate-nitrogen (NO_3-N). Efforts have been made to have nitrate analysis and tolerances for safety uniformly reported as nitrate-nitrogen on 100 percent dry matter basis. However, at present, reports may be given as nitrate or nitrate-nitrogen and may be reported as either percent or as parts per million (ppm).

Toxic Effects of Nitrate

With all the biological variation that can exist in both plants and animals, it is obviously difficult to develop specific guidelines that fit all conditions. Safe levels of nitrate are not specifically known for all the various livestock feeding conditions. At the same time, some general guidelines are needed. The concept in developing the tables that follow is to be conservative to help assure safety. At the same time, reasonable feeding and care of the animal is assumed. There are two kinds of possible toxicity that are of concern: (1) Acute or lethal and (2) Chronic or non-lethal.

Uggestions for Minimising Nitrate Problems

Toxicity or safety of a feed containing more than normal amounts of nitrate involves many factors. These include total daily intake of nitrate, previous adaptation of the animal to high nitrate, feeding practices, nutritional quality of the ration and general health of the animal. In addition, the nitrate level of the water may be a contributing factor.

The following are the major factors that influence possible nitrate problems. Preventing or changing a condition that increases the problem will help decrease or prevent the problem. Total nitrate intake is the critical problem rather than the amount in any one feed in the ration. For example, a dangerous level in a feed that makes up the total ration could be perfectly safe if it comprised only half the ration. Likewise, a safe level in the feed may be a problem if the water also contains high levels of nitrate.

Determine nitrate in all suspicious feeds and in the water. Limit the amount of any questionable feed. Forages, especially corn and oat silages, green chop or pasture and weeds, are the most common sources of the problem. Limits may need to be greater if the water also contains significant nitrate levels.

Nitrate in one dose may be very toxic while the same amount divided into several smaller doses is perfectly safe. For example, in one study, giving a 1000 pound cow 150 grams (about 5 ounces) of nitrate (NO_3) in one dose produced acute toxicity. Spraying three times as much, or 450 grams (about one pound) of nitrate on the hay consumed in a day, did not produce acute toxicity.

If a feed contains questionable amounts of nitrate, divide the daily feeding into smaller feedings. For example, feed ten pounds of silage 3 or 4 times per day rather than feeding 30 or 40 pounds at one feeding.

Nitrate is not normally accumulated in the animal because it is continually converted to other nitrogen compounds that are utilised or excreted in the urine and feces. The ability to utilise and effectively excrete the nitrogen compounds requires adaptation by the animal. A toxic level given to a cow that had been on a very low nitrate intake could be a safe level if the same amount had been gradually added to the ration. If a toxic level is fed repeatedly, liver and kidney damage can occur. If a feed is questionable, feed a small amount for a week; if no problem is noted the amount can be increased. When changing to a new feed or different source of feed, it is always better to make the change gradually. Safe utilisation of nitrate requires general good nutrition and proper rumen function in cattle, sheep and goats. Rumen micro-organisms require readily available carbohydrate, protein and minerals. Additional vitamin A reduces the toxic effects of nitrates in poultry.

Feed a balanced ration. Liberal feeding of a good grain mix insures adequate levels of energy and protein. Minerals should be provided in the grain mixture, as well as by free choice feeding of trace mineral salt and a calcium-phosphorus supplement. Nitrates might destroy or interfere with the conversion of carotene to vitamin A; therefore, the use of a protein source that contains vitamin A or adding a vitamin A supplement in some way is generally advised when high nitrate feeds are used. Avoid feeding high nitrate feeds to animals that are unthrifty or sick for any reason. Do not feed questionable forages to cows that have impaired rumen function. Elimination of high nitrate feeds from rations for dry and recently early lactation cows is also a good practice.

Nitrates can accumulate in plants when adverse growing conditions such as drought, hot weather, cool weather or frost slows the growth of the plants. Cool season crops such as small grains and permanent forage grasses may accumulate nitrates in hot, dry weather while warm season crops such as corn and sorghums can accumulate nitrates when the temperatures are low or when growth has been arrested due to frost.

Nitrates primarily accumulate in the lower stems and leaves of corn, sorghums, small grains, grasses and weeds. They seldom accumulate at sufficiently high concentrations to be a problem in legumes. If the nitrates are suspected to be a problem, avoid harvesting the basal portions of the plant. Avoid pasturing hungry animals or green feeding on drought stunted crops or weedy pastures. Place suspect crops in the silo and allow to ferment for one to three months before feeding and then follow suggestions given previously. Ensiling will allow conversion of nitrate to ammonia and may reduce nitrate levels by 30 to 50%. Haymaking does not reduce the nitrate level of the forage.

Nitrates are hazardous to man. Silo gases are a common cause of both chronic and acute respiratory problems to farmers. When plants are high in nitrate at time of ensiling, various nitrogen gases are formed. Nitrogen dioxide is a brownish yellow gas with a bleach-like odour, that is sometimes seen coming down the silo chute or is layered on top of the silo below closed doors. The gas is heavier than air so it accumulates in low enclosed areas. Nitrogen dioxide is potentially very dangerous to man and animals.

Don't leave the silo room door open to the barn as the gas may come down from the silo and kill animals in the barn overnight. Be sure to ventilate a recently filled silo before entering it by running the blower for several minutes. Open outside doors or windows near bottom of the silo chute and use fan to force gas outside. Escaping nitrogen in the form of gas from the silo reduces nitrate content of silage. Presence of the yellow gas during filling does not mean the silage will be dangerous as a feed, however, it is good evidence that high nitrate was present. Therefore, this silage should be checked for nitrate before feeding.

Highly Pathogenic H5N1 Avian Influenza and Pets

Can Highly Pathogenic H5N1 Infect Pets?

The highly pathogenic Asian Strain of H5N1 avian influenza (AI) currently found in Asia, Africa and Europe can infect multiple species of domestic (chickens, turkeys, quails, guinea fowl, etc.) wild and pet birds. This virus has also been detected in mammalian species including humans, rats and mice, weasels and ferrets, pigs, cats and dogs.

However, the number of documented cases of AI H5N1 in non-avian species is very low despite the fact that this virus has caused large avian outbreaks globally, over the last few years.

How is it Transmitted?

It is believed that the vast majority of human cases of H5N1 Asian strain of AI are the result of direct and close contact with infected birds. There have been no documented or reported cases of AI H5N1 being transmitted to humans from mammalian pets.

To date, the only feral cats shown conclusively to be naturally infected with AI H5N1 were found in an area of Germany which experienced a significant outbreak in wild birds in February, 2006. It is suspected that these cats became infected after eating infected wild birds.

There have been additional reports of increased mortality in cats during AI H5N1 outbreaks in other countries but these have not been substantiated by laboratory tests. Some non-domestic cats (tigers and leopards) have also contracted the disease in zoos after eating raw infected poultry meat. Serological studies in several Asian countries suggest that dogs have been exposed to the virus. An unpublished study carried out in 2005 by the National Institute of Animal Health in Bangkok showed that 160 out of 629 village dogs (25%) had antibodies to AI H5N1. No clinical cases of AI H5N1 have been reported in dogs.

Can Infected Animals, Such as Cats, Transmit the Virus to Humans?

The World Organisation for Animal Health and the Food and Agriculture Organisation of the United Nations confirm the statement issued by the World Health Organisation (WHO) in February 2006 that "there is no present evidence that domestic cats can play a role in the transmission cycle of H5N1 viruses".

Current science suggests that the risk of a human being contracting AI from a mammalian pet is very low. Nonetheless, owners are encouraged to take appropriate precautions to protect their pets and themselves. Pet owners should contact their veterinarian if they have any concerns about the health of their pet.

Should Pet Owners Take Special Precautions?

In view of the susceptibility of certain types of cats, it is recommended that cats in infected zones and surveillance zones set up around AI outbreaks be kept indoors. Dogs in these areas should be kept on a leash as a precautionary measure. Pet birds, if they are not normally kept indoors, should also be restricted to the indoors in areas that are experiencing AI H5N1 outbreaks in domestic or wild birds.

The Food and Agriculture Organisation (FAO) of the United Nations provides the following advice to cat owners in areas where highly pathogenic H5N1 has been diagnosed or suspected in poultry or wild birds:

- be especially vigilant for any dead or sick cats and report such findings to the local veterinarian;
- do not touch or handle any sick looking or dead cat (or other animal) and report to the authorities;
- wash hands with water and soap regularly and especially after handling animals and cleaning their litter boxes or coming in contact with feces or saliva;
- make sure contact between cats and wild birds or poultry (or the feces) is avoided, and/or keep cats inside;
- if cats bring a sick or dead bird inside the house, put on plastic gloves and dispense of the bird in plastic bags for collection by local veterinary animal handlers;
- keep stray cats outside the house and avoid contact with them; and
- if cats show breathing problems or nasal discharge, a veterinarian should be consulted.

It must be stressed that pets should not be abandoned under any circumstances. In some countries there has been public fear and uncertainty about the health of pets, mostly cats, and their role in spreading AI H5N1. Pets have been given up and abandoned as a result of these concerns. This seriously compromises the welfare of the animal through starvation and exposure to accidental injury and disease.

Bordetella Bronchiseptica

Bordetella bronchiseptica is a small, gram-negative, rod-shaped bacterium of the genus *Bordetella*. It can cause infectious bronchitis, but rarely infects humans. Closely related to *B. pertussis*—the obligate human pathogen that causes pertussis or *whooping cough—B. bronchiseptica* can persist in the environment for extended periods.

Pathogenesis

Humans are not natural carriers of B. bronchiseptica, which typically infects the respiratory tracts of smaller mammals (cats, dogs, rabbits, etc). People are more likely to be infected by B. pertussis or B. parapertussis.

B. bronchiseptica does not express pertussis toxin, which is one of the characteristic virulence factors of B. pertussis. But it has the genes to do so, highlighting the close evolutionary relationship between the two species.

Veterinary Pathogenesis

In veterinary medicine, *B. bronchiseptica* leads to a range of pathologies in different hosts. It is a serious disease of dogs, pigs, and rabbits and has been seen in cats, horses, and seals. There is a PCR test for the pathogen.

In pigs, *B. bronchiseptica* and *Pasteurella multocida* act synergistically to cause atrophic rhinitis, a disease resulting in arrested growth and distortion of the turbinates in the nasal terminus (snout).

In dogs, *B. bronchiseptica* causes acute tracheobronchitis, which typically has a harsh, honking cough. Kennel cough can also be caused by canine adenovirus-2 or canine parainfluenza virus or a combination of pathogens.

In rabbits, *B. bronchiseptica* is often found in the nasal tract. It is often assumed to cause a nearly asymptomatic infection known as snuffles, but the causative agent for that disease is *Pasteurella multocida*; "*B. bronchiseptica*" often co-infects the nasal passage at the same time.

Bronchiseptica

Cats infected with *B. bronchiseptica* have been seen with tracheobronchitis, conjunctivitis and rhinitis (upper-respiratory tract infection or URI), mandibular lymphadenopathy, and pneumonia. However, URI in cats can also be caused by herpesvirus, calicivirus, *Mycoplasma* species, or *Chlamydia psittaci*. An intranasal vaccine exists for cats.

Prevention and Control of Bordetella Bronchiseptica Infection in Cats

Until now prevention of feline upper respiratory tract disease (URTD) has been limited to cattery management.

Particularly good hygiene practices and the reduction of stress in cats, is an important aid in the prevention of URTD in cats. Removal of chronically-infected cats from affected catteries can also be considered.

An intranasal vaccine is now available in some countries and this can be used to help prevent URTD outbreaks in catteries. Vaccination is also indicated to help prevent respiratory disease in cats which:

- come into contact with other cats and dogs
- are boarded at catteries or are exhibited.

Cat vaccination programmes should always be combined with good husbandry: good nutrition, sanitation, ventilation, parasite control and control of other respiratory pathogens to minimise the occurrence of clinical disease. During parturition, queens may pass *Bordetella bronchiseptica* to their offspring; kittens obtained from breeding catteries and rescue shelters are, therefore, particularly at risk from this disease. Since *Bordetella bronchiseptica* infection in kittens can cause acute death as a result of bronchopneumonia, this risk must be taken seriously. The potential for mortality and the general morbidity caused by infectious URTD warrants the use of measures designed to tackle disease, particularly in high-risk animals. Intranasal Vaination is an important part of such preventative schemes.

Good Cattery Practice and Prevention

Bordetella bronchiseptica (Bb) infection can be prevented by a combination of good cattery practice, careful management and vaccination. *Bb* does not survive long outside the cat and is killed by many common disinfectants so routine hygiene measures are sufficient to prevent disease spreading through the environment.

Recommendations for good cattery practice:

- clean and disinfect cages, food bowls and litter trays on a daily basis e.g. hypochlorite or quaternary ammonium compounds
- use disinfectant baths for foot-dipping in between entering pens
- enter each cage as few times as possible
- finish all cleaning and disinfection tasks in one area before moving on to the next
- disinfect cages thoroughly between cats, preferably leaving empty for 2 days
- put cats with any previous respiratory problems or suspected carrier cats at one end of the cattery and feed last
- avoid overcrowding: provide one pen per cat wherever possible (unless the cats come from the same household)
- maintain optimal environmental conditions for comfort: optimum temperature, good ventilation, low relative humidity, good drainage

Infected cats can be treated with antibiotics but blanket treatment of cats in your cattery is needed. This is expensive and carrier cats may still shed *Bb* and infect healthy cats. This form of control can encourage the development of antibiotic resistance. Good cattery practice requires veterinary advice on vaccination programmes aimed to prevent feline URTD.

Streptococcal Infection in Poultry

Species are the cause of opportunistic infections in poultry leading to acute and chronic conditions in affected birds. Disease varies according to the Streptococcal species but common presentations include septicaemia, peritonitis, salpingitis and endocarditis.

Common species affecting poultry include:

- *S. gallinaceus* in broiler chickens
- *S. gallolyticus* which is a pathogen of racing pigeons and turkey poults
- *S. dysgalactiae* in broiler chickens
- *S. mutans* in geese
- *S. pluranimalium* in broiler chickens
- *S. equi subsp. zooepidemicus* in chickens and turkeys
- *S. suis* in psittacine birds.

Diagnosis

Post-mortem findings include friable internal organs, abdominal effusion and evidence of sepsis in the joints, heart valves and brain.

Bacteria can usually be cultured from tissues collected at necropsy or identified by microscope examination.

Treatment and Control

The organism should be cultured and antibiotic sensitivity should be determined before treatment is started. Amoxycillin is usually effective in treating streptococcal infections. Biosecurity protocols and good hygiene are important in preventing the disease.

Vaccination is available against S. gallolyticus *and can also protect pigeons.*

Streptococcus & Enterococcus Infections-Poultry

Streptococcal and enterococcal infections in poultry can cause acute septicaemia and chronic infections in affected birds.

There are ~50 recognised members of the Streptococcus genus and 21 of Enterococcus spp. all are commensal organisms, primarily of the gastrointestinal tract and mucosal surfaces, in both animals and humans. Some are also ubiquitous in the environment.

The majority of infections from these pathogens are opportunistic.

There does not appear to be any zoonotic risk to humans from ingestion of infected birds with these infections, with the exception of

Streptococcus suis which is rare in birds and more commonly acquired from pigs. There are however concerns regarding the spread of Vancomycin Resistant Enterococci (VRE) to humans from poultry.

Diseases—Streptococcosis

Thus clinical signs also vary widely and may range from anorexia, lethargy and diarrhoea to high mortality rates and severe neurological signs, lameness or jaundice.

Many other species not discussed here have been isolated from healthy birds and are thought to be non-pathogenic.

Known as a facultative pathogen in racing pigeons causing acute mortality, lameness, weight loss, diarrhoea and inability to fly. Also causes mortality in turkey poults.

Associated with endocarditis, peritonitis, salpingitis, egg drop, septicaemia and high mortality in chickens and also massive mortality in eared grebes. In turkeys, the same pathogen is thought to be responsible for septicaemia.

Enterococcosis

Enterococci are frequently involved in infections of day old chicks, affecting primarily the yolk sac.

Lesions

On necropsy, changes in chickens, ducks and turkeys are similar and various organs may be affected, dependent of the species involved. Septicaemic birds are the most common presentation, in which the liver is pale and friable and viscera within both abdominal and thoracic cavities are congested with yellow or orange-ish discolouration. There may be a sero-fibrinous or serosanguinous exudate covering some organs. The skin may be discoloured and crusting in cases of cellulitis/ dermatitis and joint lesions possibly with yellow caseous exudates occur in septic arthritis. Vegetative lesions are indicative of endocarditis. Infarcts can be present in any organ. In birds with encephalomalacia, lesions may be found anywhere except the cerebellum and yellow discolourations will be seen; malacia and heterophilic infiltration are visible microscopically. Pectoral muscle necrosis is pathognomic for S. gallolyticus septicaemia in pigeons.

Diagnosis

Isolation and identification is essential for confirmed diagnosis. Bacteriology can be performed on yolk, embryo fluid, valvular vegetations from necropsy, blood samples from septicaemic birds, joint and any suspected lesions. It is thought that capillary microblood cultures may be more sensitive due to trapping of the micro-organisms in these small vessels.

The pathogens can then be cultured on sheep or ox blood agar, selective blood agar or MacConkey agar. Serological Lancefield testing can also be performed on beta haemolytic colonies.

The organism can often be visualised in microscopic sections as generic gram positive cocci.

Treatment

Culture and sensitivity should always be performed whenever possible before selecting treatment. The efficacy of treatment deteriorates with progression of disease within a flock. Chronic cases, especially those involving endocarditis and arthritis, often do not respond.

Amoxycillin appears to be the drug of first choice for most infections according to sensitivity studies in poultry.

Ampicillin, doxycycline and erythromycin are effective in pigeon outbreaks of S. gallolyticus.

Control

Good hygiene, management and housing strategies are imperative. Hygiene in the hatcher is particularly important to prevent iatrogenic transmission. It is possible to transmit these opportunistic bacteria when giving vaccinations, e.g. those for Mareks Disease. Minimisation of stress and immunosuppressive disease is also key.

Vaccines are available against S. gallolyticus *and provide a degree of clinical protection for pigeons.*

Manual of Diagnostic Tests and Vaccines for Terrestrial Animals

The *Manual of Diagnostic Tests and Vaccines for Terrestrial Animals* (*Terrestrial Manual*) aims to facilitate international trade in animals and animal products and to contribute to the improvement of animal health services world-wide. The principal target readership is laboratories carrying out veterinary diagnostic tests and surveillance, plus vaccine manufacturers and regulatory authorities in Member Countries. The objective is to provide internationally agreed diagnostic laboratory methods and requirements for the production and control of vaccines and other biological products.

This ambitious task has required the cooperation of highly renowned animal health specialists from many countries. The OIE, the World Organisation for Animal Health, is clearly the most appropriate organisation to undertake this task on a global level. The main activities of the organisation, which was established in 1924 and in 2008 comprised 172 Member Countries and Territories, are as follows:

1. To ensure transparency in the global animal disease and zoonosis situation.
2. To collect, analyse and disseminate scientific veterinary information on animal disease control methods.
3. To provide expertise and encourage international solidarity in the control of animal diseases.
4. Within its mandate under the WTO (World Trade Organisation) Agreement on Sanitary and Phytosanitary Measures (SPS Agreement), to safeguard world trade by publishing health standards for international trade in animals and animal products.
5. To improve the legal framework and resources of national Veterinary Services.
6. To provide a better guarantee of the safety of food of animal origin and to promote animal welfare through a science-based approach.

The *Terrestrial Manual*, covering infectious and parasitic diseases of mammals, birds and bees, was first published in 1989. Each successive edition has extended and updated the information provided. This sixth edition includes new chapters on Guidelines for international standards for vaccine banks, Turkey rhinotracheitis (avian metapneumovirus), Small hive beetle infestation (*Aethina tumida*) and camelpox, and *Mycoplasma synoviae* has been added to the chapter on Avian mycoplasmosis (previously the chapter focused on *Mycoplasma gallisepticum*.

As a companion volume to the *Terrestrial Animal Health Code*, the *Terrestrial Manual* sets laboratory standards for all OIE listed diseases as well as several other diseases of global importance. In particular it specifies (in blue font) those “Prescribed Tests” that are recommended for use in health screening for international trade or movement of animals. The *Terrestrial Manual* has become widely adopted as a key reference book for veterinary laboratories around the world. Aquatic animal diseases are included in a separate *Aquatic Manual*.

The task of commissioning chapters and compiling the *Terrestrial Manual* was assigned to the OIE Biological Standards Commission by the International Committee of the OIE (General Assembly of national Delegates of Member Countries and Territories). Manuscripts were requested from specialists in each of the diseases or the other topics covered. After initial scrutiny by the Consultant Technical Editor, the chapters were sent to scientific reviewers and to experts at OIE Reference Laboratories. They were also circulated to all OIE Member Countries for review and comment. The Biological Standards Commission and the Consultant Technical Editor took all the resulting comments into consideration, often referring back to the authors for further help, before finalising the chapters. The final text has the approval of the International Committee of the OIE.

A procedure for the official recognition of commercialised diagnostic tests, under the authority of the International Committee, was finalised in September 2004. Data are submitted using a validation template that was developed by the Biological Standards Commission. Submissions are evaluated by appointed experts, who advise the Biological Standards Commission before the final opinion of the OIE International Committee is sought. All information on the submission of applications can be found on the OIE Web site.

Arrangement of the Terrestrial Manual

Terrestrial Manual, contains eleven introductory chapters that deal with a variety of general subjects of interest to veterinary laboratory diagnosticians. These chapters are intended to give a brief introduction to their subjects. They are to be regarded as background information rather than standards.

The main part of the *Terrestrial Manual* (Part 2) covers standards for diagnostic tests and vaccines for the diseases listed in the OIE *Terrestrial Animal Health Code.* The diseases are in alphabetical order, subdivided by animal host species. OIE listed diseases are transmissible diseases that have the potential for very serious and rapid spread, irrespective of national borders. They have particularly serious socio-economic or public health consequences and are of major importance in the international trade of animals and animal products. Four of the diseases in Section 2.9 are included in some individual species sections, but these chapters cover several species and thus give a broader description. Some additional diseases that may also be of importance to trade but that do not have a chapter in the *Terrestrial Code* are also included in Section 2.9. The contributors of all the chapters are listed on pages xxii–xxxv, but the final responsibility for the content of the *Terrestrial Manual* lies with the International Committee of the OIE.

Format of Chapters

Each disease chapter includes a summary intended to provide information for veterinary officials and other readers who need a general overview of the tests and vaccines available for the disease. This is followed by a text giving greater detail for laboratory workers. In each disease chapter, Part A gives a general introduction to the disease, Part B deals with laboratory diagnosis of the disease, and Part C (where appropriate) with the requirements for vaccines or *in vivo* diagnostic biologicals.

The information concerning production and control of vaccines or diagnostics is given as an example; it is not always necessary to follow these when there are scientifically justifiable reasons for using alternative approaches. Bibliographic references that provide further information are listed at the end of each chapter.

Explanation of the Tests Described and of the Table on Pages xi–xiv

The table on pages xi–xiv lists diagnostic tests in two categories: 'prescribed' and 'alternative'. Prescribed tests are those that are required by the *Terrestrial Animal Health Code* for the testing of animals before they are moved internationally. In the *Terrestrial Manual* these tests are printed in blue. At present it is not possible to have prescribed tests for every listed disease. 'Alternative tests' are those that are suitable for the diagnosis of disease within a local setting, and can also be used in the import/export of animals after bilateral agreement. There are often other tests described in the chapters, which may also be of some practical value in local situations or which may still be under development.

World Animal Health Information

The WAHID Interface provides access to all data held within OIE's new World Animal Health Information System (WAHIS). It replaces and significantly extends the former web interface named Handistatus II System.

A comprehensive range of information is available from:

- Immediate notifications and follow-up reports submitted by Member Countries in response to exceptional disease events occurring in these countries as well as follow-up reports about these events,
- Six-monthly reports describing the OIE-listed disease situations in each country

- Annual reports providing further background information on animal health, on laboratory and vaccine production facilities, etc.

To start, select one of the headings on the left. You can then explore available information

- by country (or group of countries),
- by disease,
- focusing on control measures, or
- comparing the animal health situation between two countries.

Veterinary Medicine

Intute: Veterinary medicine provides free access to high quality resources on the Internet. Each resource has been evaluated and categorised by subject specialists based at UK universities.

No new resources are being added to the catalogue, but existing resources will be checked and broken links will be fixed until July 2011.

Chapter 9

Animal Health Strategy

A Challenging EU Animal Health Strategy

Based on the evaluation results and the stakeholder consultation, the Commission is pleased to present its proposal for a new EU Animal Health Strategy (2007-2013).

This will allow further debate in the EU inter-institutional fora, with Council and the Parliament expected to establish their positions by the end of this year. Overall, the strategy encompasses a challenging 6 year programme of work aimed towards clear outcomes:

- Prioritisation of EU intervention;
- A modern and appropriate animal health framework;
- Better prevention, surveillance and crisis preparedness;
- Science, Innovation and Research.
- The timetable for delivery of all the specific actions included in this strategy will depend on the position of the Council and the Parliament, and also on our human resources capacity.

Vision

Our vision is to work in partnership to increase the prevention of animal health related problems before they happen: "Prevention is better than cure".

Purpose

The strategy provides direction for the development of animal health policy, based on extensive stakeholder consultation and a firm commitment to high standards of animal health. It will facilitate the establishment of priorities that are consistent with agreed strategic goals and the revision of, and agreement on, acceptable and appropriate standards.

Scope of the Strategy

The concept of animal health covers not only the absence of disease in animals, but also the critical relationship between the health of animals and their welfare. It is also a pillar for the Commission's policy on public health and food safety.

The strategy covers the health of all animals in the EU kept for food, farming, sport, companionship, entertainment and in zoos. It also covers wild animals and animals used in research where there is a risk of them transmitting disease to other animals or to humans. The strategy also covers the health of animals transported to, from and within the EU. The strategy is aimed at the entire EU, including animal owners, the veterinary profession, food chain businesses, animal health industries, animal interest groups, researchers and teachers, governing bodies of sport and recreational organisations, educational facilities, consumers, travellers, competent authorities of Member States and the EU Institutions.

The strategy builds on the current animal health legal framework1 in the EU and the standards and guidelines of the World Organisation for Animal Health (Office international des Epizooties – OIE). It will aim at ensuring consistency with other EU policies and the EU's international commitments.

It will guide the development of new policies or guidelines and will enhance existing animal health arrangements in the Community based on scientific risk assessments and taking into account social, economic and ethical considerations. It will support the achievement of a high level of environmental protection by considering the impacts on the environment in the development of the policy framework.

Goals

The strategy sets out some challenging aims, not just for the EU institutions and Governments, but for all citizens, to improve animal health. The strategy's goals are:Goal 1

- To ensure a high level of public health and food safety by minimising the incidence of biological and chemical risks to humans.Goal
- To promote animal health by preventing/reducing the incidence of animal diseases, and in this way to support farming and the rural economy.Goal

- To improve economic growth/cohesion/competitiveness assuring free circulation of goods and proportionate animal movements.Goal
- To promote farming practices and animal welfare which prevent animal health related threats and minimise environmental impacts in support of the EU Sustainable Development Strategy.

Simple and reliable performance indicators will help to measure progress towards the strategy's goals, guide policy, inform priorities, target resources and focus discussion. They will be developed in consultation with stakeholders and improved over time as better veterinary and other data becomes available. They will cover both hard indicators of animal health (e.g. disease prevalence, number of animals eliminated) and softer indicators tracking the confidence, expectations and perceptions of European citizens. It must be recognised that uncertainties and unforeseeable events may affect achievement of the performance indicators..

The action plan will aim to explain the significant breadth of activity which is being carried out or will be carried out at EU level through legislative proposal or other mechanisms in order to deliver the different strategy goals over the next six years.

The action plan to deliver the strategic goals (section 4) will focus on four main pillars, or areas of activity:

A partnership approach built on trust, openness and a willingness to take difficult decisions is essential for success. The strategy can only bring about real change if everyone involved in animal health works together and with all interested citizens. There are many excellent examples of partnership in action in the current Community Animal Health Policy. We must take advantage of existing collaborations, encourage new initiatives and make more use of non-legislative alternatives to regulation.

An "Animal Health Advisory Committee" will include representatives from non governmental organisations across the animal health sector, consumers and governments. The "Animal Health Advisory Committee" will provide strategic guidance on the appropriate/acceptable level of animal or public health protection, and on priorities for action and communication. It will also follow the strategy's progress: it will be consulted on all impact assessments and will advise the Commission on the best means of delivering agreed outcomes.

CommUnIC AtIon

Animal health is a concern for all European citizens. This concern stems from the public health and food safety aspects of animal health but also from the economic costs that animal disease outbreaks can trigger and the animal welfare considerations, including the implications of disease control. The Commission is committed to pursue its objectives of clarity and transparency when communicating with consumers and stakeholders what the EU is doing and why. European and national entities need to cooperate to ensure a coherent message and enhance public confidence.

Communicating the Strategy

There will be annual reporting on the strategy's progress and wider communication of policies and initiatives. Communication will take different forms depending on the message that is being delivered and the target audience. It will include participation in international or national events, developing relationships with the media and non governmental organisations.

Communication in Case of Crisis

Good communication on risk to stakeholders/consumers is also of utmost importance, as an incorrect public perception of risk may force the regulator to take unjustified or disproportionate measures in the case of a crisis. The Animal Health Advisory Committee will advise the Commission to further improve communication during a crisis situation. A new Animal Health Strategy for the European Union (2007-2013) where "Prevention is better than cure"all interested parties in the risk management process to gain the widest possible agreement and shared responsibility for the judgements made and to deliver agreed objectives.

Targets will be set at community level, national level and, where appropriate, regional level. Suitable performance indicators will allow the assessment of progress over the next six years. The appropriate amount of resources to be applied to achieve the desired level of protection, and the development of a responsibility and cost sharing scheme, will be based on the categorisation of biological and chemical risks.

Categorisation of biological and chemical risks according to level of relevance for the EU;

Agreement on the acceptable level of risk;

Setting of priorities, quantifiable targets and performance indicators;

Setting of the amount of resources to be committed to identified threats.

A new Animal Health Strategy for the European Union (2007-2013) where "Prevention is better than cure"

Disease can be devastating on farmers and the economy as a whole – in a specific country, a continent or even globally.

International organisations such as OIE and the World Bank consider Animal Health as a *global public good*. The EU considers the maintenance of "Animal Health Services" in line with international standards (in terms of legislation, structure, organisation, resources, capacities, the role of the private sector and paraprofessionals) as a minimum goal. This is a public investment priority.

Constantly evolving legislation is one of the main mechanisms for EU intervention in animal health, both in the pursuit of Community policy and to implement international obligations. Better regulation principles will be applied through a strengthened partnership and enhanced communication.

The future strategy will aim at replacing the existing series of linked and interrelated policy actions by a single policy framework. The Animal Health Strategy will strive for a single clear regulatory framework converging as far as possible with the OIE/Codex recommendations/standards and guidelines. This regulatory framework will include animal nutrition and animal welfare measures.

The European Commission is responsible for ensuring that unjustified national/ regional rules do not constitute a potential obstacle to the internal market. However, the EU regulatory framework also needs to be suitably flexible to allow for judgements of equivalence, settlements of dispute and efficient responses to changing situations.

Specific attention must be paid to the position of animals kept on a non-commercial basis (i.e. as a hobby) and wildlife, insofar as this impinges on central goals.

Roles and responsibilities will be clearly defined. An incentive-oriented approach is needed at all levels. A revision of the current co-financing instrument is required. More effective procedures will be used for a number of Commission Decisions. The SCFCAH9 will focus on Decisions in which MS and stakeholders have a key interest. Non-regulatory tools must be encouraged as far as possible.

A single horizontal legal framework '! will define and integrate common principles and requirements of existing legislation (intracommunity trade, imports, animal disease control, animal nutrition and animal welfare).

Existing legislation will be '! simplified and replaced by this new framework as appropriate, seeking convergence to international standards (OIE/Codex standards) while ensuring a firm commitment to high standards of animal health.

Animal Diseases

Existing compensation schemes are mainly focused on providing a compensation mechanism for animal owners in the event of a disease outbreak. Appropriate sharing of costs, benefits and responsibilities could contribute significantly to the key objectives of the strategy. It could contribute to preventing major financial risks for Member States and the Community by providing incentives for prevention of animal related threats. It would also seek to strengthen Community economic and social cohesion and specifically to reduce the gaps between levels of animal health in the various regions.

On the one hand, Governments have an important role to play in securing our external borders against disease incursions and leading the response to outbreaks of exotic disease. Provision of state compensation is also of utmost importance to compensate for private property destroyed for the public good at least to the extent that the owner is not responsible for the outbreak. In this the protection of public health is a key consideration. On the other hand, responsibility for the health of animals lies primarily with animal owners and collectively with the industry. As a result, animal owners and industry are better placed than others to deal with many of the risks of animal diseases.

There is a clear recognition that the policy needs the full participation and commitment of all parties, including the insurance sector. Ownership of risk is a key issue and new mechanisms must be introduced to involve major stakeholders in decision-making on significant policy issues, in particular for emergency measures. A feasibility study will be necessary to reflect on concrete proposals for the gradual development of an EU harmonised scheme.

Feed Sector

In the feed sector, when large-scale incidents occur, public authorities tend to be heavily burdened with the costs of withdrawal, transport, storage and destruction of feed, food and animals, as well as the costs of analysis and other administrative outlay. Feed business operators are liable for any infringements of the relevant legislation on feed safety and for the direct consequences of the withdrawal from the market, treatment and/ or destruction of any feed, animals and food produced there from. In 2007, the Commission will submit a report to the European Parliament and the Council setting out the possibilities for an effective system of financial guarantees for feed business operators.

The high level of animal health within the EU will make a key contribution to growth and jobs in Europe by ensuring that farmers and European companies remain competitive and that they have genuine access to the export markets. We need to ensure that European companies, often small and medium-sized enterprises (SMEs), are able to compete fairly in those markets. Unjustified sanitary barriers tend to be increasingly important. They are complicated, technically challenging and time consuming to detect, analyse and remove. Import conditions for food of animal origin and animal products are largely harmonised. However, this is not the case for exports.

The Commission has exclusive competence for negotiation of bilateral agreements with third countries in the SPS field. For certain third countries, common EU export requirements are specifically defined in bilateral veterinary agreements. Ongoing trade negotiations, in particular the negotiations for Free Trade Agreements with Korea, India and ASEAN include SPS chapters.

The Commission is in a discussion with Member States on the implications of the implementation of existing and future policy on SPS negotiations with third countries in relation to exports. The aim is to ensure respect of the Treaty obligations in relation to the Common Commercial Policy and to present a unified Community approach in negotiations with third countries. The new EU Animal health strategy will contribute to adapt the mix of policy organisations, where appropriate.

Successful biosecurity measures must address isolation of new animals brought to the farm, isolation of sick animals, regulation of the movement of people, animals, and equipment, correct use of feed, and procedures for cleaning and disinfecting facilities. This responsibility lies with the animal owners, including hobby farmers. However, as some contagious pathogens may easily spread from one farm to another a collective approach must be taken in addressing prevention and biosecurity measures.

Effective on-farm biosecurity measures will constitute an important criterion of zoning and compartmentalisation procedures for disease control and/or trade purposes. The EU is the biggest food importer in the world. The Member States' responsibility in border control is to protect the community from potential animal and public health risks arising from international trade of live animals and their products. The challenge is to improve border biosecurity without severely disrupting cross-border movement of people and agricultural goods. In fact, the main safety feature of border controls on declared imports for animal health purposes is the document check, and the EU is dependent upon the accuracy and honesty of the declarations in these documents.

Veterinarians need to work more closely with customs, both at border inspection posts and at points of entry to the Community where goods or animals may enter illegally. There are fundamental questions about: the assessment of risk; trust between national governments; and what can and cannot be accomplished in border inspection facilities and other points of entry (efficiency/effectiveness). On the other hand, it may be difficult for certain developing countries to comply with EU standards and thus engage in trade. The EU should build on ongoing initiatives and improve cooperation with third countries, providing them with technical assistance to help them to satisfy EU animal health requirements for imports and to fight against exotic diseases at source.

Emergency Preparedness

Animal-related emergencies must be dealt with swiftly and effectively using an agreed approach. The possibility for the Commission to take fast-track decisions for emergency action is of high value in limiting and controlling animal-related threats at EU level. In response to ethical concerns and the growing demand for improved animal welfare, the EU has already moved to a more flexible approach to vaccination, as well as improving its policy to control major animal diseases. To decrease the number of animals eliminated will be one of the objectives of the new EU animal health policy (goal 4).

However, different elements make it important that the decision to use vaccination is taken on a case by case basis. Preparation, contingency planning exercises and implementation of emergency preparedness plans are the responsibility of the governments. These plans should be agreed in advance with the cost sharing scheme partners. A key factor in being able to manage an outbreak successfully is knowing where animals and their products are, and controlling their movements.

Animal Testing

The history of animal testing goes back to the writings of the Greeks in the 4th and 3rd centuries BCE, with Aristotle (384-322 BCE) and Erasistratus (304-258 BCE) among the first to perform experiments on living animals. Galen, a physician in 2nd-century Rome, dissected pigs and goats, and is known as the “father of vivisection.” Avenzoar, an Arabic physician in 12th-century Moorish Spain who also practiced dissection, introduced animal testing as an experimental method of testing surgical procedures before applying them to human patients.

Basic Science Advances

In 1242, Ibn al-Nafis provided accurate descriptions of the circulation of blood in mammals. A more complete description of this circulation was later provided in the 17th century by William Harvey. In the 18th century, Antoine Lavoisier, used a guinea pig in a calorimetre to prove that respiration was a form of combustion, and Stephen Hales measured blood pressure in the horse. In the 1780s, Luigi Galvani demonstrated that electricity applied to a dead, dissected, frog's leg muscle caused it to twitch, which led to an appreciation for the relationship between electricity and animation. In the 1880s, Louis Pasteur convincingly demonstrated the germ theory of medicine by giving anthrax to sheep. In the 1890s, Ivan Pavlov famously used dogs to describe classical conditioning.

In 1921 Otto Loewi provided the first strong evidence that neuronal communication with target cells occurred via chemical synapses. He extracted two hearts from frogs and left them beating in an ionic bath. He stimulated the attached Vagus nerve of the first heart, and observed its beating slowed. When the second heart was placed in the ionic bath of the first, it also slowed.

In the 1920s, Edgar Adrian formulated the theory of neural communication that the frequency of action potentials, and not the size of the action potentials, was the basis for communicating the magnitude of the signal. His work was performed in an isolated frog nerve-muscle preparation. Adrian was awarded a Nobel Prize for his work.

In the 1960s David Hubel and Torsten Wiesel demonstrated the macrocolumnar organisation of visual areas in cats and monkeys, and provided physiological evidence for the critical period for the development of disparity sensitivity in vision (i.e.: the main cue for depth perception), and were awarded a Nobel Prize for their work.

In 1996 Dolly the sheep was born, the first mammal to be cloned from an adult cell.

Medical Advances

In the 1880s and 1890s, Emil von Behring isolated the diphtheria toxin and demonstrated its effects in guinea pigs. He went on to demonstrate immunity against diphtheria in animals in 1898 by injecting a mix of toxin and antitoxin. This work constituted in part the rationale for awarding von Behring the 1901 Nobel Prize in

Physiology and Medicine. Roughly 15 years later, Behring announced such a mix suitable for human immunity which largely banished the diphtheria from the scourges of mankind. The antitoxin is famously commemorated each year in the Iditarod race, which is modelled after the delivery of diphtheria antitoxin to Nome in the 1925 serum run to Nome. The success of the animal studies in producing the diphtheria antitoxin are attributed by some as a cause in the decline of the early 20th century antivivisectionist movement in the USA.

In 1921, Frederick Banting tied up the pancreatic ducts of dogs, and discovered that the isolates of pancreatic secretion could be used to keep dogs with diabetes alive. He followed up these experiments with chemical isolation of insulin in 1922 with John Macleod. These experiments used bovine sources instead of dogs to improve the supply. The first person treated was Leonard Thompson, a 14 year old diabetic who only weighed 65 pounds and was about to slip into a coma and die. After the first dose, the formulation had to be reworked, a process that took 12 days. The second dose was effective. These two won the Nobel Prize in Physiology or Medicine in 1923 for their discovery of insulin and its treatment of diabetes mellitus. Thompson lived 13 more years taking insulin. Before insulin's clinical use, a diagnosis of diabetes mellitus meant death; Thompson had been diagnosed in 1919.

In the 1943, Selman Waksman's laboratory discovered streptomycin using a series of screens to find antibacterial substances from the soil. Waksman coined the term antibiotic with regards to these substances. Waksman would win the Nobel Prize in Medicine in 1952 for his discoveries in antibiotics. Corwin Hinshaw and William Feldman took the streptomycin samples and cured tuberculosis in four guinea pigs with it. Hinshaw followed these studies with human trials that provided a dramatic advance in the ability to stop and reverse the progression of tuberculosis. Mortality from tuberculosis in the UK has diminished from the early 20th century due to better hygiene and improved living standards, but from the moment antibiotics were introduced, the fall became much steeper, so that by the 1980s mortality in developed countries was effectively zero.

In the 1940s, Jonas Salk used Rhesus monkey cross-contamination studies to isolate the three forms of the polio virus that affected hundreds of thousands yearly. Salk's team created a vaccine against the strains of polio in cell cultures of Rhesus monkey kidney cells. The vaccine was made publicly available in 1955, and reduced the incidence of polio 15-fold in the USA over the following five years.

Albert Sabin made a superior "live" vaccine by passing the polio virus through animal hosts, including monkeys. The vaccine was produced for mass consumption in 1963 and is still in use today. It had virtually eradicated polio in the USA by 1965. It has been estimated that 100,000 Rhesus monkeys were killed in the course of developing the polio vaccines, and 65 doses of vaccine were produced for each monkey.

Also in the 1940s, John Cade tested lithium salts in guinea pigs in a search for pharmaceuticals with anticonvulsant properties. The animals seemed calmer in their mood. He then tested lithium on himself, before using it to treat recurrent mania. The introduction of lithium revolutionized the treatment of manic-depressives by the 1970s. Prior to Cade's animal testing, manic-depressives were treated with lobotomy or electro-convulsive therapy.

In the 1950s the first safer, non-volatile anaesthetic halothane was developed through studies on rodents, rabbits, dogs, cats and monkeys. This paved the way for a whole new generation of modern general anaesthetics-also developed by animal studies-without which modern, complex surgical operations would be virtually impossible.

In 1960, Albert Starr pioneered heart valve replacement surgery in humans after a series of surgical advances in dogs. He received the Lasker Medical Award in 2007 for his efforts, along with Alain Carpentier. In 1968 Carpentier made heart valve replacements from the heart valves of pigs, which are pre-treated with gluteraldehyde to blunt immune response. Over 300,000 people receive heart valve replacements derived from Starr and Carpentier's designs annually. Carpentier said of Starr's initial advances "Before his prosthetic, patients with valvular disease would die".

In the 1970s, leprosy multidrug antibiotic treatments were refined using lerosy bacteria grown in armadillos, and were then tested in human clinical trials. Today, the nine-banded armadillo is still used to culture the bacteria that causes leprosy, for studies of the proteomics and genomics (the genome was completed in 1998) of the bacteria, for the purposes of improving therapy and developing vaccines. Leprosy is still prevalent in Brazil, Madagascar, Mozambique, Tanzania, India and Nepal, with over 400,000 cases at the beginning of 2004. The bacteria has not yet been cultured in vitro with success necessary to develop drug treatments or vaccines, and mice and armadillos have been the sources of the bacteria for research.

The non-human primate models of AIDS, using HIV-2, SHIV, and SIV in macaques, have been used as a complement to ongoing research efforts against the virus. The drug tenofovir has had its efficacy and toxicology evaluated in macaques, and found long-term/high-dose treatments had adverse effects not found using short-term/high-dose treatment followed by long-term/low-dose treatment. This finding in macaques was translated into human dosing regimens. Prophylactic treatment with anti-virals has been evaluated in macaques, because introduction of the virus can only be controlled in an animal model. The finding that prophylaxis can be effective at blocking infection has altered the treatment for occupational exposures, such as needle exposures.

Such exposures are now followed rapidly with anti-HIV drugs, and this practice has resulted in measurable transient virus infection similar to the NHP model. Similarly, the mother-to-fetus transmission, and its fetal prophylaxis with antivirals such as tenofovir and AZT, has been evaluated in controlled testing in macaques not possible in humans, and this knowledge has guided antiviral treatment in pregnant mothers with HIV. "The comparison and correlation of results obtained in monkey and human studies is leading to a growing validation and recognition of the relevance of the animal model. Although each animal model has its limitations, carefully designed drug studies in nonhuman primates can continue to advance our scientific knowledge and guide future clinical trials."

Throughout the 20th century, research that used live animals has led to many other medical advances and treatments for human diseases, such as: organ transplant techniques and anti-transplant rejection medications, the heart-lung machine, antibiotics like penicillin, and whooping cough vaccine.

Presently, animal experimentation continues to be used in research that aims to solve medical problems from Alzheimer's disease, multiple sclerosis spinal cord injury, and many more conditions in which there is no useful in vitro model system available.

Treatments to each of the following animal diseases have been derived from animal studies: rabies, anthrax, glanders, Feline immunodeficiency virus (FIV), tuberculosis, Texas cattle fever, Classical swine fever (hog cholera), Heartworm and other parasitic infections.

Basic and applied research in veterinary medicine continues in varied topics, such as searching for improved treatments and vaccines for feline leukemia virus and improving veterinary oncology.

Early Debate

In 1655, physiologist Edmund O'Meara is recorded as saying that "the miserable torture of vivisection places the body in an unnatural state." O'Meara thus expressed one of the chief scientific objections to vivisection: that the pain that the subject endured would interfere with the accuracy of the results.

In 1822, the first animal protection law was enacted in the British parliament, followed by the Cruelty to Animals Act (1876), the first law specifically aimed at regulating animal testing. The legislation was promoted by Charles Darwin, who wrote to Ray Lankester in March 1871:

You ask about my opinion on vivisection. I quite agree that it is justifiable for real investigations on physiology; but not for mere damnable and detestable curiosity. It is a subject which makes me sick with horror, so I will not say another word about it, else I shall not sleep tonight."

Opposition to the use of animals in medical research arose in the United States during the 1860s, when Henry Bergh founded the American Society for the Prevention of Cruelty to Animals (ASPCA), with America's first specifically anti-vivisection organisation being the American AntiVivisection Society (AAVS), founded in 1883.

In the UK, an article in the Medical Times and Gazette on April 28, 1877, indicates that anti-vivisectionist campaigners, mainly clergymen, had prepared a number of posters entitled, "This is vivisection," "This is a living dog," and "This is a living rabbit," depicting animals in a poses that they said copied the work of Elias von Cyon in St. Petersburg, though the article says the images differ from the originals. It states that no more than 10 or a dozen men were actively involved in animal testing on living animals in the UK at that time.

Antivivisectionists of the era generally believed the spread of mercy was the great cause of civilization, and vivisection was cruel. However, in the U.S., the antivivisectionists' efforts were defeated in every legislature, overwhelmed by the superior organisation and influence of the medical community.

The early antivivisectionist movement in the U.S. dwindled greatly in the 1920s, potentially caused by a variety of factors including opposition of the medical community, improvement in medicine through the use of animals, and the tendency of the antivivisectionists to misrepresentation and exaggeration, and their use of inaccurate, vague and outdated references. Overall, this movement had no US legislative success until the passing of the Laboratory Animal Welfare Act, in 1966.

On the other side of the debate, those in favour of animal testing held that experiments on animals were necessary to advance medical and biological knowledge. The founders, in 1831, of the Dublin Zoo—

the fourth oldest zoo in Europe, after Vienna, Paris, and London—were members of the medical profession, interested in studying the animals both while they were alive and when they were dead.

Claude Bernard, known as the "prince of vivisectors" and the father of physiology—whose wife, Marie Françoise Martin, founded the first anti-vivisection society in France in 1883—famously wrote in 1865 that "the science of life is a superb and dazzlingly lighted hall which may be reached only by passing through a long and ghastly kitchen." Arguing that "experiments on animals...are entirely conclusive for the toxicology and hygiene of man...the effects of these substances are the same on man as on animals, save for differences in degree," Bernard established animal experimentation as part of the standard scientific method.

In 1896, the physiologist and physician Dr. Walter B. Cannon said "The antivivisectionists are the second of the two types Theodore Roosevelt described when he said, 'Common sense without conscience may lead to crime, but conscience without common sense may lead to folly, which is the handmaiden of crime.'"

These divisions between pro-and anti-animal testing groups first came to public attention during the brown dog affair in the early 20th century, when hundreds of medical students clashed with anti-vivisectionists and police over a memorial to a vivisected dog.

Is Climate Change Increasing Viral Disease in Farm Animals?

Since 1998, there have been increasing outbreaks of the viral disease, bluetongue, among European livestock. Using a newly developed climate-based model that accurately predicts past outbreaks, researchers have provided evidence that climate change is a major driver of these outbreaks.

Bluetongue (BT) is a viral disease of ruminants (which include sheep and cattle) spread by biting midges. For some time it has been widely distributed in Africa, Asia, Australia, South America and North America. Until recently, it was very rare in Europe, but in 1998, an unprecedented series of outbreaks began, causing the deaths of millions of animals and serious economic impacts.

Scientists have proposed that changes in climate may be partly responsible. Furthermore, since it has emerged only recently, scientists believe BT provides a good example for the development of models to describe how climate change may influence diseases in the future.

The study, partly supported by the EU CIRCE project1, developed a framework to evaluate the effects of past and future climate on the risk of the emergence of BT.

The framework integrated climate data into a disease transmission model and evaluated whether the model could reproduce accurately past outbreaks of BT. Good correspondence between the model and historic records suggests that climate was influencing the increase in outbreaks.

Sheep and cattle were considered as these are present in large numbers; data on midge abundance was provided by the Spanish BT national surveillance programme. Past climate data from 1961-2008 was derived from a dataset of observed temperature and rainfall, and future climate data from model simulations provided by ENSEMBLES2 data. The climate-driven model explained both spatial and temporal variations in the recent emergence and spread of BT in Europe. This included the 2006 BT outbreak in northwest Europe, which occurred in the year that the model predicted the highest risk since at least 1960.

The model also provided insight into the role of drivers in different parts of Europe. For example, in northwest Europe temperature is a major driver of disease risk, whereas in southwest Europe the situation is more complex. In southwest Europe, if rainfall is low then increases in temperature will lead to a decrease in risk of BT outbreaks, and if rainfall increases, higher temperatures will lead to an increase in risk.

The model also quantified the effects of future changes of climate on the risk of BT transmission. Results suggest that by 2050 the number of secondary cases arising from the introduction of one infected animal will increase by 30 per cent in northwest Europe and by 10 per cent in southwest Europe.

The researchers highlighted some weaknesses in the model, such as the lack of species-specific data of the different types of midges, including biting rate, incubation period and mortality rate. They also suggested the data on midge abundance was questionable as it relied on midges caught in light traps, which might not represent the population. Lastly, there was no accounting for lags in the model which presumes a near-instant impact of changes in climate. Nevertheless, the framework successfully describes past outbreaks of BT in Europe, provides predictions for the future and could be applied to other diseases.

Emerging Animal Diseases and International Trade

Infectious animal and zoonotic diseases are becoming more and more important in the changing farming and trading systems of the industrialised and developing world, writes TheCattleSite Senior Editor, Chris Harris.

Dr. Alejandro A. Schudel from Fundación PROSAIA in Argentina told the World Meat Congress in Cape Town, that some emerging or evolving infectious diseases have the potential to quickly spread from local to international significance, or to jump species barriers-including to humans.

He added that these diseases can have a serious impact on international trade and also in the perception of the consumers with regard to the safety of animal derived products.

"We are facing new challenges in animal diseases," Dr Schudel said.

"In the last 25 years they have spread very quickly and have taken on a global perspective."

He said that there is a significant interaction between both human and animal health now that we face animal and zoonotic diseases of unprecedented magnitude. He said the world is at the threshold of a new era of emergent and re-emergent diseases that will have a tremendous impact on public health. The significance of its potential consequences has substantially influenced the acceptance of the concept of "Healthy Animals, Safe Food, Healthy People."

Examples of Emerging and Re-Emerging Diseases

Dr Schudel said that this is reflected in the fact that numerous countries have established their own standards and regulations to control diseases. The effects of globalisation, industrialisation, the restructuring of the agricultural sector, and the growing and sustained demand for animal proteins, are rapidly changing the foundations and the applications of policies directly related to animal health and food safety as well as to the safety of the food supply.

One important aspect is that the increase in diseases is largely in zoonotic diseases-those that can be passed from animal to human.

He estimated that 75 per cent of the emerging diseases are zoonotic. The factors that influence the spread of disease are environmental, ecological, genetic and political. Poverty and inequality, wars, a lack of domestic political will to tackle disease all influence how it is spread around the world.

Dr Schudel added that this new scenario of worldwide health affects not only the importing and exporting countries, but all countries as a whole, due to the tremendous implications that such events could have on global trade. The recent outbreaks of Bovine Spongiform Encephalopathy (BSE) in Europe, Asia and North America, the Highly Pathogenic Avian Influenza (HPAI) epidemic that started in South East Asia and the introduction of Foot and Mouth (FMD) disease in Europe, are typical examples of this new scenario. The quick detection of these emerging, re-emerging and zoonotic diseases and the consequent reaction will be the decisive factor in order to be able to quickly overcome the threat posed by such events.

He said that the lessons learned from the outbreaks of these diseases were that there was an increased need for food safety measures, increased communication to the consumer and there was a raised awareness among the general public of issues of animal welfare and well being. He added that the impact of the diseases was felt by everyone-rich and poor alike and that there was a serious financial consequence to any outbreak of disease. The 2001 UK foot and mouth disease outbreak had a financial impact of £13 billion. In the US it has been estimated that the financial costs of an outbreak of foot and mouth disease would be between $7 billion and $14 billion. The cost of the avian influenza outbreak in Asia has already been estimated at between $8 billion and $12 billion.

Animal health has a lot to do with food safety and healthy animals have a lot to do with safe food he said. The readiness of any country to face an animal emerging disease, and its capacity to respond, depend in great measure to the preparedness of its Veterinary Services.

The lack of infrastructures which support diagnostic laboratories, veterinarians, expert services and worldwide vigilance, are insufficient, and this fact is clearly seen among developed and developing nations.

He said there is an urgent need to improve veterinary services and a need for transparent veterinary services. Another extremely important aspect that needs to be addressed is the quick and efficient manner on how to alert the consumer about the scientific based risks of these diseases with regards to food safety.

He said that the outbreaks of BSE and Avian Influenza are cases that show the need for transparency and swift communication.

Animal Health authorities and Public Health authorities have to confront the issues in a coordinated manner. International organisations are bringing about joint efforts, and the World Organisation for Animal Health (OIE) has clearly said it should play a role when facing the difficulties that these diseases pose on a worldwide level.

Effective global security can only be achieved if all trading countries comply with the internationally agreed upon standards and guidelines, effectively training people in the industry and ensuring the availability of adequate human and material veterinary resources and he also called for a strong global quality certification process to verify the integrity of products and the health of animals. Many countries share a common concern about the emergence of animal diseases and zoonosis or the deliberate misuse of pathogenic biological agents that could affect the public health as well as food and animal production.

Existing methods of disease prevention and containment regulations as well as international guidelines and standards are available and should be followed in order to procure early detection and a quick containment, Dr Schudel said.

Predicting Changes for Parasite and Vector Induced Animal Diseases

Climate change could have a major influence on animal health, both directly and indirectly, by affecting the parasites and vectors that spread diseases, according to Eric Hoberg, an Agricultural Research Service (ARS) zoologist at the Agricultural Research Centre in Beltsville. Climate change can alter an animal's relationship with parasites and vectors. These changes can influence where parasites and vectors thrive, making certain geographical regions more or less amenable to them. Climate change can also alter when and for how long parasites and vectors pose a threat to agricultural animals. Climate can determine how pathogens are distributed, transmitted and evolve, and can influence the factors associated with emerging disease and how animals respond to those diseases. Significant environmental changes have been well documented in recent decades, and some of these changes are causing trouble for livestock.

Dr Eric Hoberg is one of many ARS scientists investigating the probable impact of climate changes on agricultural parasites and virus vectors. Their research is generating information that could help producers prepare for and respond to heightened disease threats.

In one study, Dr Hoberg collaborated with several Canadian scientists to investigate the influence of climate change on parasitic lungworms known as *Protostrongylus stilesi* in Arctic mammals. Although the lungworms had never before been observed in muskoxen, the scientists observed them in muskoxen that share habitat with Dall's sheep. Environmental changes that bring the two animals into contact more frequently could result in larger parasite populations.

"Climate change alters the boundaries between different species and between natural and managed lands," Dr Hoberg says. "When these boundaries break down, it becomes possible for pathogens to switch between hosts."

Muskoxen are not a major agricultural animal in the United States, but their interaction with lungworms and Dall's sheep could be used to establish predictive models that could help US animal producers better understand how pathogens switch between hosts. Climate change may also influence insects that spread diseases. At the ARS Arthropod-Borne Animal Diseases Research Laboratory (ABADRL) in Laramie, Wyoming, scientists are investigating several vectored diseases, including bluetongue and Rift Valley fever (RVF).

"In recent years we've seen an incursion of bluetongue virus in Europe, but whether that's affected by climate change or not hasn't yet been determined," says ABADRL microbiologist, Bill Wilson.

Muskoxen

Several factors influenced by climate change could explain the spread of bluetongue virus, says ABADRL entomologist, Kristine Bennett. Higher temperatures could enable *Culicoides imicola,* the disease's primary vector in Europe, to inhabit a much larger geographic area. Higher temperatures could enhance viral replication, potentially enabling indigenous Culicoides populations to vector the disease. And warmer winter temperatures could enable more Culicoides to survive the winter, extending their influence. Other factors are at play in the case of RVF, a viral disease that occurs in Africa and recently has spread beyond the continent's borders into the Arabian Peninsula.

"RVF outbreaks are closely connected to cycling global climate variability caused by El Niño Southern Oscillation phenomena – particularly rainfall and flooding, which introduce the virus into livestock populations by facilitating the hatching of virus-infected mosquito eggs resting in the soil," says Ken Linthicum, director of the ARS Centre for Medical, Agricultural and Veterinary Entomology in Gainesville, Florida.

"Our research hasn't focused on climate trends, but we do know that elevated sea surface temperatures, such as those associated with El Niño cycles, are related to heavy rainfall and flooding, which are related to RVF outbreaks," Dr Linthicum says.

Dr Linthicum led a team of scientists in developing a model that in 2006 successfully predicted an RVF outbreak in Africa several months before the outbreak occurred. This early warning enabled international aid organisations to increase disease surveillance and to conduct public information, animal vaccine and insect control programs. Since then, the model has predicted outbreaks in southern Africa, Sudan and the Horn of Africa. Tools like this could be essential for early detection and control of RVF, should it ever enter the United States. Since the 1950s, ocean buoys and satellite instrument monitors have clearly documented increasing surface temperatures in the Indian Ocean. Dr Linthicum and his colleagues believe these rising temperatures and related rainfall result in more frequent, smaller outbreaks of RVF.

"When large outbreaks do occur, they tend to be bigger and more widespread than those we observed in the past," he says. This information could help policymakers mobilise outbreak response efforts.

"The potential effects of climate change are not yet fully understood," says ARS national programme leader, Cyril Gay. "But evidence suggests that animal health could be affected by rising ambient and ocean temperatures. To protect US livestock from parasite and vector-borne diseases, scientists must be engaged in research on many fronts."

Regular surveillance will provide up-to-date information about changes in pathogen populations. Laboratory and field research will help illuminate how climate changes influence pathogen characteristics, and models will help researchers and producers predict and plan for pathogen threats. With a nationwide network of research facilities and a history of groundbreaking pathogen research, ARS is well positioned to provide research, expertise and tools to address any parasite or vector-borne disease threats that arise in response to a changing global climate.

Animal Welfare Guidelines for Beef Cattle Farms

There is a significant body of national and EU legislation enacted relating to animal welfare. Here, the Irish Agriculture and Food Development Authority explain how compliance with welfare legislation is an important requirement for the National Beef Assurance Scheme and the Code of Good Farming Practice.

Many retailers and major food service chains will only purchase Irish beef that originates from assured farms. Animal welfare standards are a prominent feature of Irish and European farm assurance schemes.

The welfare of farm animals is an increasingly important issue for the Irish beef industry. Research studies conducted by the Teagasc National Food Centre (NFC) and various other researchers indicate that both domestic and foreign consumers of Irish beef are becoming more conscious of animal welfare issues. The NFC studies suggest that consumers may also view high animal welfare standards as an indicator that food is safe, healthier and of high quality. Thus, perception of animal welfare standards can affect the consumers' image of Irish beef.

Good animal welfare has always being an integral part of the husbandry content of Irish livestock production systems. Irish beef production systems are grass based and extensive by nature. Nonetheless, there are aspects of beef cattle production such as the housing of animals during winter, castration, dehorning, transport, handling and slaughter that have the potential to cause stress, pain and injury if not managed correctly.

This booklet briefly summarises key animal welfare standards and best farm assurance practice that producers should be able to demonstrate to other stakeholders in the food chain. The appendices detail sources of relevant animal welfare legislation.

The Five Freedoms Concept

In essence an animal welfare Code of Practice is the application of sensible and sensitive animal husbandry practices to the livestock present on the farm. Animal welfare is concerned with the well-being of the animal and complements the objectives of beef assurance schemes that demonstrate the production of safe beef to consumers and food chain stakeholders.

Welfare codes usually list five basic freedoms that should underpin on farm animal welfare best practice. The five freedoms are listed below and provides an overall concept of animal welfare. One-freedom from thirst, hunger and malnutrition Two-freedom from discomfort Three-freedom from pain, injury and disease Four-freedom to express normal patterns of behaviour Five-freedom from fear and distress

Welfare Guidelines

Stockman

Stockmanship is a key factor in animal welfare. The stockman should have training and or the necessary experience in cattle husbandry. Without competent diligent stockmanship animal welfare will be compromised.

A competent stockman should be able to:

- recognise whether or not the animals are in good health (signs of ill health include: loss of appetite, listlessness, cessation of cudding, discharge from eyes or nostrils, dribbling, persistent coughing, lameness, swollen joints, scouring, rapid loss of condition or emaciation, excessive scratching, abnormal skin conditions or other unusual conditions)
- understand the significance of a change in the behaviour of the animals

- know when veterinary treatment is required
- implement a planned herd health programme (e.g. preventative treatments, vaccination programmes if necessary)
- implement appropriate animal feeding and grassland management programmes
- recognise if the general environment (indoors or outdoors) is adequate for the promotion of good health and welfare
- have management skills appropriate to the scale and technical requirements of the production system
- handle animals with care, avoiding undue stress.

A good stockman will individually inspect all animals at least once per day. Particular categories of animals will require more frequent inspection e.g. young calves or cows in late pregnancy etc. Formal training and/or experience working under the supervision of a competent stockman is strongly recommended where inexperienced persons are taking over responsibility for animal husbandry on a farm.

Common veterinary type activities (e.g. dosing, injecting, castration) should not be attempted without direct appropriate supervision until the stockman is competent to carry out these activities. People already involved in animal management/husbandry should keep themselves updated in technological developments that can prevent or correct welfare problems.

Husbandry Practices and Relevant Records

Husbandry practices should minimise stress to the animal. All farms must have proper animal handling facilities including pens and a crush where an animal can be restrained with minimum risk of injury or stress. Good handling facilities also benefit the safety of the personnel involved in handling the animals. Early and frequent contact with competent persons particularly at an early age greatly reduces the stress to animals subsequently. Cattle are gregarious animals who will socialise with each other, when young calves are individually penned they should be able to see other calves.

Many assurance schemes and regulatory requirements require that key records are maintained on the farm. Some of these records help producers to demonstrate that best practice has been implemented in relation to animal health and welfare standards. Key records include:

- Bovine Herd Register
- Animal Remedies Record
- Animal Feed Records.

Some assurance schemes also require producers to maintain a Planned Herd Health Programme Checklist. Some retailer assurance schemes also require producers to document recommended operating procedures in the event of an emergency (fire in a livestock shed, operational guidelines for replacement stockperson in the event that the stockperson is away)

Basic Farm Biosecurity

Producers are advised to follow good on-farm biosecurity measures to protect their livestock and crops from the constant threat of pests and diseases. Biosecurity Queensland principal veterinary officer Janet Berry said being aware of biosecurity meant keeping animals safe from disease and ensuring continued market access for produce.

"Straightforward measures built into everyday practice will go a long way toward protecting your farm and your future," she said.

"Animal owners should assess the risks to their animals and act to reduce the risks."

Purchased Livestock

The movement of new animals onto your property represents the highest risk of introducing disease into your herd or flock. Inspect the animals carefully for disease before buying them. Always request the history and supporting paperwork, such as the vendor declaration or national health statement, before you buy the animals. Isolate new animals to make sure they are disease and weed free before mixing with your stock.

Stray Animals

Poor fencing can allow stray livestock, and wild or feral animals to mix with your stock and introduce disease. Keep all gates shut and check fences regularly.

People

People can carry animal pests and diseases. Ask all visitors where they have been previously; whether they've had contact with other animals, or been abroad and possibly brought diseases home. Keep a register of all visitors. Restrict visitor access to your property and make sure they don't go near animals unless they have clean clothes and have disinfected hands and footwear.

Vehicles and Equipment

Vehicles and equipment can carry pests and diseases. Control the entry of vehicles onto the property and ensure they stay in a designated vehicle area. Use your own vehicles to transport visitors or material around the farm. Maintain clean and disinfected equipment and do not share with other animal owners.

Feed and Water

Feed and water can contain pests and diseases. Always request a commodity vendor declaration with purchased feed. Keep feed in a clean dry storage area and ensure it does not become mouldy. Make sure that water sources are not contaminated by wild or feral animals or birds.

Dr Berry said there were major outcomes of having farm biosecurity plans in place.

"Outcomes include improved profitability through the reduction of diseases, less need for expensive chemical treatments or vaccinations and improved animal production," she said.

"Farm biosecurity plans can reduce the risk of introducing pests and diseases onto a property that are already present on your neighbours farm or elsewhere.

"Good planning now will also reduce the impact of the next disease emergency.

"The rapid and wide geographical spread of emergency diseases can be controlled more easily if all livestock owners begin to practice farm biosecurity now."

Global Approaches to Farm Animal Welfare

This article is taken from the proceedings of the Livestock Care Conference, held by Alberta Farm Animal Care. New emphasis on animal welfare in the U.S., Europe and internationally is changing the dynamics of how food is produced and marketed worldwide.

For those still not convinced that consumer concerns over animal welfare represent anything more than a fringe mentality, consider the following: according to a 2007 American Farm Bureau survey, 68 percent of Americans polled think the government should take an active role in promoting the welfare of farm animals.

Nothing new, right? Most producers have heard statistics like this before, but there are two things that set this one apart. First, it's the result of a poll conducted by an independent grassroots farm agency rather than a group, such as People for the Ethical Treatment of Animals (PETA), with a specific agenda on animal welfare. Secondly, it's a call fr government intervention in a country that typically waves the flag for government that is as small as possible.

To Dr. Ed Pajor, director for the Centre for Animal Well-Being at Purdue University in Indiana, it's an example of a new social ethic concerning animals – and one that has turned long-relied-upon models of livestock production and marketing on their heads.

Pajor spent two years studying the animal welfare culture of Europe and the United States – two markets with far different approaches to animal welfare that are nevertheless heading toward a similar end-point. In a presentation at the Livestock Care Conference in Red Deer, Alberta April 4, he offered an overview of his findings and what they may mean to the Canadian livestock industry going forward. "Emerging animal welfare policy from powerful world bodies such as the World Organisation for Animal Health (OIE) is set to have a major influence on policy in North America," he says. "Meanwhile, the free market is already having a major impact, with major food companies increasingly preferring to work with suppliers who can provide proof of responsible animal welfare practices.

"This means higher expectations for the livestock industry, but it can also mean a great opportunity for those who can provide this assurance. The bottom line is that demand for transparent animal welfare will not be going away. It's a reflection of a whole new way in which society views animals, and is quickly becoming part of the culture of agriculture as well."

The OIE Animal-centred Approach

No discussion of farm animal welfare at the global level would be complete without mentioning the World Organisation for Animal Health, otherwise known as the OIE. Created in 1924 to develop standards to combat the outbreak of animal diseases, OIE standards have become the international reference in the field of animal diseases for the World Trade Organisation (WTO). In recent years, this relationship has helped the OIE become a key developer of international standards for animal welfare as well.

Countries are under no obligation to adopt OIE standards and guidelines, says Pajor. However, the OIE's emphasis on science-based guidelines and its "five freedoms" of animal care (freedom from hunger and thirst; freedom from discomfort; freedom from pain, injury and disease; freedom to express normal behaviour; and freedom from fear and distress) have made the organisation a standard-bearer for countries seeking approaches that satisfy customer demands.

"What sets the OIE's guidelines apart is its emphasis on animal-based criteria as opposed to resource-based criteria. The difference is that the former sets out easily-measurable guidelines related to design and inputs such as space allowances, temperature ranges, and air quality. Animal-based criteria, on the other hand, focuses on performance and output factors such as survival rate, disease and injury, behaviour, and reaction to handlers. These criteria are very difficult to measure, but they address the needs of individual animals much more comprehensively."

Speaking in a general sense, Europe is different culturally from countries such as the U.S. in that its citizens look to government for a comprehensive degree of leadership, says Pajor. And, true to the wishes of its populace, Europe took a lead on animal welfare in the 1960s, when the Council of Europe began to focus on animal welfare issues.

Today, animal welfare has become part of Europe's culture of agriculture. The European model of developing animal welfare law follows the top-down model of the OIE – albeit on a more prescriptive basis – by providing guidelines and minimum requirements that often become law at the national level.

The Council of Europe, which includes 46 member states, issues both binding conventions and non-binding guidelines on animal care for its members. The 27 countries that make up the European Union also fall under directives established by the EU-27, he says. These directives, which are almost always based on reports from scientific experts, generally focus on animal-centred outcomes such as space allowance per animal, freedom of movement, social interaction and the limitation of painful interventions.

In addition to government-driven legislation and standards, animal welfare also plays a large role in industry-based quality assurance schemes. "Industry-driven quality assurance is considered very important to European consumers, especially niche markets, and are often stricter than other regulations," says Pajor.

Although much of the above may sound very prescriptive, the fact is that Europe has based most of its animal welfare activity on the wishes of European consumers. They are able to do so because of a strong foundation of data measuring public opinion – a tool that is largely missing in North America, says Pajor.

"The Eurobarometre, a series of surveys on the attitudes of Europeans on a variety of topics, breaks opinion on animal welfare issues down by individual countries. Not surprisingly, these opinions vary in the same way they might vary between, for example, Calgary and Toronto. The overall message, however, is that, to Europeans as a whole, animal welfare is considered an important attribute of overall food quality."

This infrastructure of public opinion is helping to drive the Welfare Quality Project, a $27 million effort involving 17 countries and 44 universities and institute which stands to become the next generation of animal welfare in Europe. One of the key goals of the project is to develop criteria which capture the public's description of animal welfare.

"I believe this project will have a significant impact on North American agriculture," says Pajor. "It won't reflect just a few people's opinion – it will be the opinion of a lot of people in a lot of countries with a lot of scientific power backing it up."

Animal Welfare: Driven by Economics

In the U.S., it's been the private sector that has taken the lead on animal welfare activity, with Burger King's decision to discontinue the acceptance of hogs raised in gestation stalls being one of the most prominent examples. "For the most part, however, animal welfare in the U.S. is based on voluntary guidelines, although state bans on specific production systems are appearing and animal law is becoming a rapidly growing area of law," says Pajor.

Animal welfare is being built into several private sector-based quality assurance strategies in the U.S. However, Pajor says there is a general tendency to emphasize resource or engineering-based criteria over the kind of animal-focused criteria modelled by the OIE. This approach may not be enough moving forward, however.

"A recent Gallup poll showed that 71 percent of respondents felt that animals deserved some protection, while 25 percent felt that they deserved the same rights as people," says Pajor. "Twenty-five percent is a huge number. And while it's possible that many respondents had companion animals in mind when forming these results, there are implications here on how they think about other types of animals as well. The fact that the same poll showed that 62 percent of respondents favoured passing strict laws concerning the treatment of farm animals bears this out."

Pajor believes consumers, often informed by activists, will continue to play a role in industry action on animal welfare. "By targeting just a few firms, activists are achieving de facto changes in industry regulations in the U.S. These regulations are not imposed by legislators, but by firms seeking a competitive advantage. If you can convince Wal-Mart that animals should be produced in a certain way, there will be a huge impact on how they're produced."

Where We're Headed?

The bottom line, says Pajor, is that animal welfare is here to stay, with guidelines and standards continuing to be developed at all levels, including international agreements and multinational companies. A key strategy moving forward, he says, will be to bullet-proof animal welfare assurance programs against external criticism as much as possible.

"These processes will need to be as transparent as possible, complete with third party audits. Having codes without assessment, audit and teeth is something that will come under more and more criticism."

The livestock industry can also expect animal welfare expectations to become more animal-centred, he says. "We will have more and more undercover video of operations that have been certified. If assurance programs are not taken seriously, consumers will lose trust in our systems and the public has limited trust already. If animal welfare is not embraced wholeheartedly as a necessary element of animal production, grassroots efforts will fall apart and be filled by legislation."

Animal Science-Recent Progress and Future Challenges

The goal of animal science remains efficient animal production according to Professor John Oldham, reports editor, Jackie Linden, for The Cattle Site.

Professor John Oldham of Scottish Agricultural Colleges in Edinburgh was invited to give the Hammond Lecture by the British Society of Animal Science (BSAS) at its annual meeting in Belfast this year. He discussed the developments and understanding bought by previous generations of animal scientists, including Sir John Hammond, in whose memory the opening lecture of the BSAS is dedicated, and how these studies remain just as important today to feed a growing world population at a time of climate change.

Animal science spans a broad compass, Professor Oldham said. It tends to be considered as a study of animals (nature, function, productivity etc) that are under the management of Man in some way. This differentiates it from zoology, excludes wildlife and provides boundaries with at last some aspects of veterinary and biomedical sciences that can be rather indistinct. It covers both the understanding of animals as biological systems and the role of animals as components of broader systems.

"By being allied to animals that are under management, I take it as implicit that any consideration of progress in animal science needs to include not only progress in intellectual understanding but also progress in the utility of that understanding in practice. This does not deny the value of animals to animal science to inform broader biological understanding," said Professor Oldham.

There has been considerable scientific effort and increases in understanding, both on animals themselves and on their roles in systems, although rather more in the former than in the latter, he explained. Equally, there are future challenges both at the level of the animal as a phenotype and for animals in systems of management, In each case, the great challenge is not simply to understand the behaviour – in the broadest sense – of the animal or system, but also to predict it.

'Recent' is a word that is open to some interpretation. The rate at which some new techniques cna help to generate data in research can make it appear that rates of progress are accelerating. 'Recent' progress in such areas might be considered to span small numbers of years. But a broader perspective is needed to recognise some of the major areas of progress.

Progress in any science is driven by a mixture of imagination to generate the ideas that are worth pursuing, and the acquisition of appropriate techniques to allow those ideas to be tested. Sir John Hammond was a fount of ideas on reproduction, growth and development in the inter-war years and through the 1950s. His insights and techniques led the way in practical improvements in reproduction and his concepts of growth and development have generally stood the tests of time. Like Hammond's waves of growth, progress in animal science has been influenced by waves of technical development. Professor Oldham cited the eamples of the advent of chromatographic techniques in the early post-war years allowed access to rapid analytical methods that underpinned progress in nutrition and physiology especially.

Isotropic tracing techniques gave access to more quantitative understanding of metabolism as well as facilitating analysis. More recently, molecular techniques have opened avenues to understanding that were previously closed. And advance in computational and statistical techniques have not only supported the rapid delivery of thoer methods but have also allowed the realisation of ideas that, in some areas – especially genetics – were formed a significant time ago but were inaccessible for want of computational power.

Progress in genetics has helped to develop animals whose rates of productivity are dramatically increased although sometimes with negative associated consequences – which are now being corrected by adjustments to selection approaches, Professor Oldham said. The promise of molecular approaches to replace more conventional quantitative methods has not yet been fully realised, although the prospects of using genome-wide selection are considerable. Understanding of nutrition and nutritional biochemistry has yielded rationing schemes that work tolerably well and allow some control over the qualities of products that animals produce. Our understanding of reproductive processes has grown substantially, but too often the application of new knowledge has been in the 'catch-up' mode, aiming to correct problems introduced by other 'advances'.

Research on pregnancy and lactation for all mammalian species, including our own, has though, been considerably advanced through the animal sciences. Our understanding of the processes of disease have been enabled through the animal as well as the veterinary and meical sciences, and increasingly, methods of control can be expected to include broad-based approaches that will call on the products of a range of animal sciences. The welfare of animals, as sentient beings, is better appreciated and open to assessment and improvement. But the prediction of phenotypic expression – what an animal will actually do given a knowledge of its genotype and its environment – is still as aspiration rather than a reality.

Professor Oldham said that, on broader knowledge fronts, the impacts of animals on their environment are better understood now than they were. Management of waste to reduce pollution or grazing to achieve biodiversity goals is more possible. Animals also influence the social or operating environments. For example, systems of animal management can be important fro social cohesion and are important sources of work power or equity. The impacts of animals on the environment are a source of much current concern, though, he said. Some of this relates to issues around climate change, others to the management of pollution or of managing biodiversity. Yet other concerns are to do with perceived negative contributions of animal products to the healthiness of the human diet, At the same time, the growing size of the human population and its ability to afford moe animal products in the global diet has increased, and it likely to continue to increase, the global demands for animals and their products.

This is the big driver that creates the grand challenges for animal sciences to address, Professor Oldham emphasised. The need is to produce animal products in appropriate abundance that are beneficial (or at worst, not detrimental) to the healthiness of our diets, that are produced with minimal waste, (by efficient use of resources and minimisation of losses through disease and reproductive failure especially), no negative impacts on, and preferably benefits to, the environment and in socially acceptable ways.

He said: "To achieve these results, if such is possible, I would expect to see the next stages of progress in the animal sciences being more integrative and predictive. Synthesis of understanding to enable the prediction of functional behaviour of animals as systems and animals in systems (and the behaviour of those systems) as well as continuing lines of discovery to enable this to happen should be to the fore."

A sharp eye on the outcome, and not simply the understanding, is merited. Sir John Hammond was an outstanding scientist – but his goal as science to enable efficient animal production, and we still need that, Professor Oldham concluded.

Learning & Profiting from Animal Welfare Trends

England's largest supermarket paying dairy farmers 15 to 20 percent more to improve welfare. McDonald's leading the way with higher livestock welfare standards. Farmers worldwide taking steps to integrate welfare into quality assurance schemes.

These developments are signals of monumental shifts in expectations and approaches surrounding animal welfare, each with major implications for the livestock industry.

Helping make sense of the trends and what they mean for producers both in Europe and North America was Dr. John Webster of the University of Bristol in England, who delivered a presentation on the subject at the Livestock Care Conference, April 4 in Red Deer. The conference was hosted by Alberta Farm Animal Care (AFAC), a partnership of Alberta's major livestock groups with a mandate to promote responsible, humane animal care within the livestock industry.

"We have reached a pivotal time in animal welfare," says Webster. "This includes many changes and challenges. But for the livestock industry, the good news is that it also includes profit-generating opportunities for farmers related to producing a value added product, which are critical to their survival in a competitive environment."

This assessment and the details that support it mean a lot coming from Webster. One of the world's leading authorities on animal welfare issues, Webster was a founding member of the UK Farm Animal Welfare Council and is the original proponent of the "five freedoms" concept that has become central to emerging animal welfare policy and strategies worldwide. While he sees that animal agriculture still has substantial room to improve when it comes to implementing systems of continual welfare improvement, strong progress is being made and innovations are gaining steam in many countries, particularly in Europe but also in North America.

"I wouldn't want to fool anyone by saying all steps to improve animal welfare will improve profitability for livestock producers," says

Webster. "But the way things are moving, I would say to the livestock industry that being self protective and saying there's nothing wrong is not your best solution. There is an opportunity there in terms of value-added products. Just think: if we can do it better, they're very likely to pay us."

What is Welfare Science?

New Definitions, New Expectations

Critical to understanding the new trends in animal welfare is understanding the concept of welfare itself, says Webster. This is important not only to society in general and to the livestock industry, but also to the growing ranks of 'welfare scientists' who will play an essential role in the ongoing development of welfare standards.

In this context, the traditional questions and concepts related to welfare have undergone a major updating, he argues.

For example, the typical questions traditionally posed by welfare scientists to assess animal welfare are today inadequate. These include questions such as: Is the animal living a normal life? Is the animal fit and healthy? How does the animal feel?

"If the first question was adequate, we would be saying all swine should live in the woods. If the second was adequate, we'd be fine with gestation crates so long as the animal was healthy. The third question is probably the closest to being useful, but is still not quite refined enough to provide a comprehensive assessment of welfare."

For a modern definition, Webster supports one derived and extended from earlier definitions by Dr. David Fraser of the University of British Columbia and Donald Broom of Cambridge University. This definition states that animal welfare is "The physical and mental state of a sentient animal (one sensitive in perception or and feelings) as it seeks to cope with environmental challenge."

"Welfare then doesn't just mean feeling good," says Webster. "It is the full spectrum from feeling very good to genuine suffering. And it is defined by both its physical and emotional state."

The concept of sentience is key to the definition, he notes. In pending European Community legislation, farm animals are recognised as sentient creatures.

"This reinforces that farm animals are not simply commodities. They are sentient creatures and we must respect their sentience." Webster makes a distinction between animal welfare and animal well-being. Animal welfare, he says, covers the full spectrum of animal feeling. Animal well-being, on the other hand is defined by an ability to have sustained mental and physical health, to avoid suffering and to generally feel good about life.

"For farm animals, this would include things like comfort, companionship and security, which are particularly important to animals in a herd species," says Webster.

The aspiration for those looking after farm animals, he suggests, should be to promote a sense of well-being within the context of a healthy agricultural industry. This aspiration may not be practical to achieve all the time, but the goal should be to realise it "as much as possible" and to continually strive for improvement

A Sentient View of the World

Feelings that Matter

Sentience itself too, requires an updated definition in the context of animal welfare, says Webster, suggesting that a useful definition is "feelings that matter."

A sentient animal will interpret stimuli at three levels, he explains. These include a basic reflex level, a second emotional level achieved by all sentient animals and a third cognitive level achieved by humans and to some degree by most sentient animals. Sentient animals do not just live in the present, he says. They are motivated, by emotion and cognition, to action. "The strength of motivation to action is a measure of how important these things are to them."

This concept is a foundation for an increasing number of new studies in animal welfare, which study animal responses to stressors and other stimuli as a means of validating issues of animal pain and suffering and also of gauging their relative importance to animal welfare and well-being.

Earlier pioneering studies in this vein have shown, for example, that animals will self-medicate and perform similar coping actions in response to various negative stimuli.

"If the coping action is successful, the sentient animal with cognitive capacity will learn from it and repeat it under the same circumstance," says Webster. "This is evidence of a highly sentient animal.

The Cost of Coping

From an animal science point of view, Webster suggests the best studies today are those that recognise the "cost of coping" for an animal rather than simply an alarm response, since some animals may not respond greatly if they have learned the "cost" of doing so is not worthwhile or not effective. "It's much more useful to measure the stress-specific cost of coping."

This leads to a definition of suffering, notes Webster, which he defines as a state where an animal cannot cope, or has extreme difficulty coping with a stressor. "This occurs when the stressor is too prolonged or simply too much, or because the animal is unable to respond in a way that will improve how it feels." Examples he says are swine in gestation stalls or hens in battery cages, which are unable to respond in a way that will improve their condition.

Awareness of this is key to rcognising welfare issues and seeking improvements, he says.

"The old way of thinking is that animals can experience pain, but that is just about it. Today we increasingly recognise that suffering is much more than just pain and that animals have this broader capacity to suffer." They also have strong capacity for a higher quality of life, to which those responsible for animals must aspire.

The bottom line of the new definitions and thinking on welfare is that expectations are rising, says Webster. "Suffering and pleasure are defined by the capacity to feel, not the capacity to think. And we cannot assume that the capacity to feel is proportional to intelligence, as we define it." This brings animals on more of a level plane to humans in terms of what society considers acceptable in terms of being subject to pain and suffering.

Marketplace Driving New Systems

For the livestock industry, this means expectations for how farm animals are treated are also continuing to rise. "We recognise that farm animals are sentient and have capacity to suffer," says Webster. "This raises the bar for our responsibility in taking care of these animals."

While legislation for improving welfare has been relatively slow to reflect these new expectations, the free market, operating trough the likes of McDonald's and other major food retailers is having a major impact.

As a result, animal welfare is increasingly being viewed as an important component of emerging quality assurance schemes. "Increasingly and inevitably in Europe now, quality assurance schemes are including criteria for assessment of animal welfare," says Webster. "And if animal welfare is not up to the standards of the scheme, the farmer will be lost from the scheme. Many of these schemes are driven by supermarkets and big retailers who are marketing their products on a high welfare platform because consumers are making choices based on that platform."

Tuberculosis

The prevalence of bovine tuberculosis (TB) in the United States has decreased greatly since eradication efforts were first introduced in 1917. Unfortunately, the persistence of the infection in large dairy herds, imported cattle and wild animals has prevented its complete elimination.

In a biosafety level 3 facility at NADC, veterinary immunologist Ray Waters, veterinary pathologist Mitchell Palmer and molecular immunologist Tyler Thacker have evaluated the efficacy of existing vaccines and improved diagnostic assays, contributing to a significant reduction in the prevalence of TB in recent decades.

Many of their efforts target the deer—both wild and captive—that transmit it to cattle. In one project, the team collaborated with Chembio Diagnostic Systems, Inc. (Medford, N.Y.) to develop a rapid diagnostic test for the detection of TB in deer. The test uses a whole-blood sample and takes 10-15 minutes to process. The team also worked with Prionics AG (Zurich, Switzerland) to develop improved antigens to use in TB diagnostic assays. By improving existing tools and developing new control measures, the NADC scientists have helped reduce the spread of bovine tuberculosis throughout the United States.

Rift Valley Fever

Rift Valley Fever (RVF) is a potentially fatal insect-borne disease of humans and animals, such as cattle, sheep, camels and goats. Abortions and loss of young animals are associated with RVF outbreaks. The risk of severe disease and death is much higher for livestock than for humans, but the symptoms—which include fever, aching muscles and vomiting—are unpleasant for both.

Scientists at the Arthropod-Borne Animal Diseases Research Laboratory (ABADRL) at Laramie, Wyo., are working with domestic and international collaborators to protect our nation's agricultural animals against the disease. Although RVF has not yet been introduced into the United States, it is considered a significant disease threat to U.S. livestock. The ABADRL team is cooperating with the Department of Homeland Security to assess potential vaccines for the RVF virus. In addition, the scientists are developing and evaluating diagnostic tests for RVF virus antigens.

In 2006, a team established by entomologist Kenneth J. Linthicum, director of the ARS Centre for Medical, Agricultural and Veterinary Entomology in Gainesville, Fla., produced a model that successfully predicted a Rift Valley fever outbreak in sub-Saharan Africa several months in advance. The early warning allowed organisations to advise at-risk countries to increase surveillance and insect control.

Bovine Spongiform Encephalopathy

Bovine spongiform encephalopathy (BSE, or "mad cow disease") is a deadly neurological disorder characterised by abnormally folded prions—proteins that occur naturally in mammals. In humans, the related encephalopathy, caused by the ingestion of BSE-contaminated bovine products, is known as variant Creutzfeldt-Jakob disease. In both cattle and humans, the disease causes neurological dysfunction and, eventually, death.

The United States has had only three confirmed cases of BSE. NADC scientists confirmed the first case in December 2003, and ARS scientists at the U.S. Meat Animal Research Centre in Clay Centre, Neb., traced its origin to Canada. Between June 2004 and July 2006, over 759,000 cattle have been tested for BSE, two of which tested positive. Laboratory tests showed that the manifestation of BSE in the second and third cases, which originated in the United States, differed significantly from the first. Veterinary medical officer Juergen Richt and NADC colleagues helped identify and describe the atypical cases.

In related work, Richt and his colleagues evaluated eight "prion protein knock-out" Holstein calves developed in collaboration with Hematech, Inc., a pharmaceutical research company based in Sioux Falls, S.D. These were the first cattle ever produced to lack the prion protein involved in the pathogenesis of BSE. Brain material derived from these animals was not able to support BSE-amplification when tested in an in vitro BSE amplification assay. The scientists determined that the health and growth rate of these unique animals, which will be invaluable for future BSE studies, compared to those of normal Holsteins.

Zoonotic diseases are one of several challenges facing U.S. livestock, but thanks to these and other research efforts in the ARS Animal Health National Program, our agricultural animals are among the healthiest in the world.

Bibliography

Archana Satarkar: *Food Science and Nutrition*, ABD Pub, Delhi, 2008.

Arora, N. : *Manual of Animal Nutrition*, International Book, 2004.

Aruna T. Kumar : *Handbook of Animal Husbandry*, Indian Council of Agricultural Research, 2008.

Balram Pani: *Textbook of Animal Chemistry*, I K International, Delhi, 2007.

Bhosale, Dinesh T. : *Handbook of Poultry Nutrition*, International Book, 2004.

Brock, J. : *A Natural History of Domesticated Animals*, Cambridge Univ. Pr., New York, 1999.

Brown, R. E.: *Social Odours in Animals Reproduction*, Clarendon Press, Oxford, 1985.

Chandra, Rajesh : *Diseases of Poultry and Their Control*, IBDCO, Delhi, 2001.

Clark, S. and S. Lyster: *Animals and Their Moral Standing*, Routledge, London, 1997.

Cox, N. A. : *Relationship Between Aerobic Bacteria, Salmonella, And Campylobacter on Broiler Carcasses*, Journal of Food Protection, 1997.

DeGrazia, David: *Taking Animals Seriously: Mental Life and Moral Status*, Cambridge, New York, 1996.

Donna Hill : *Biosecurity in the Poultry Industry*, International Book, 2006.

Edwards, L. : *Baking for Health*, New York, U.S.A: Avery Publishing Group, 1988.

Erasmus, U. : *Fats and Oils*, Vancouver: Alive Press, 1987.

Fox, Michael Allen: *The Case for Animal Experimentation*, University of California Press, Berkeley, CA, 1986.

Gates, P.: *Animal Communication*, Cambridge University Press, Cambridge, 1997.

Guenter, W.: *Egg Nutrition and Biotechnology*, CABI Publishing, NY, 2000.

Hambidge, Gove : *Diseases and Parasites of Poultry*, Biotech Books, Delhi, 2004.

Homer O. Stuart: *Commercial Poultry Farming*, Biotech Books, Delhi, 2011.

Jacobson, M. : *Safe Food: Eating Wisely in a Risky World,* Washington, DC: Living Planet Press, 1991.

Keith Wilson N.D.P: *A Handbook of Poultry Practice*, Agrobios, Delhi, 2000.

Mandal, A.B. : *Nutrition and Disease Management of Poultry,* International Book Distributing Co, Delhi, 2004.

Mathialagan, P : *Textbook of Animal Husbandry and Livestock Extension,* International Book Distributing Co, Delhi, 2005.

Mohiuddin, S M : *Moulds and Mycotoxins in Poultry Diseases*, International Book, Delhi, 2007.

Nyholt, D.H. : *The Vitamin & Herb Guide,* Alberta, Canada: Global Health Ltd., 1992.

Owen, W Powell : *Poultry Farming and Keeping*, Biotech Books, Delhi, 2005.

Powell., C. : *Microbiological and Hydraulic Evaluation of Immersion Chilling for Poultry*, Journal of Food Protection, 1995.

Randall, C.J.: *Color Atlas of Diseases and Disorders of the Domestic Fowl and Turkey,* Iowa State University Press, UK, 1991.

Renaville, R and A Burny : *Biotechnology in Animal Husbandry*, Springer Pub, 2008.

Rogers, Katherine: *The Cat and the Human Imagination,* The University of Michigan Press, Ann Arbor, 2001.

Sainsbury, D.: *Poultry Health and Management,* Blackwell Science, US, 2000.

Sharpe, Robert: *Science on Trial: The Human Cost of Animal Experiments*, Awareness Books, Sheffield, UK, 1994.

Singh, Ram Prakash : *Modern Livestock and Poultry Production*, Biotech Books, Delhi, 2008.

Snaydon, R.W. : *Managed Grasslands,* Amsterdam & London: Elsevier, 1987.

Tabor, Roger K.: *The Wild Life of the Domestic Cat,* Arrow Books, London, 1983.

Thornhill, Nancy W.: *The Natural History of Inbreeding and Outbreeding,* Chicago Press, 1993.

Thyagarajan, D. : *Diseases of Poultry*, Satish Serial pub, Delhi, 2011.

Ucko, Peter J. and G.W. Dimbleby: *The Domestication and Exploitation of Plants and Animals*, Aldine Publishing Company, Chicago, IL, 1969.

Watson, R.: *Eggs and Health Promotion,* Iowa State Press, UK, 2002.

Index

❑❑❑